MW00811746

Palgrave Studies in Cultural and Intellectual History

Series Editors

The Palgrave Studies in Cultural and Intellectual History series has three primary aims: to close divides between intellectual and cultural approaches, thus bringing them into mutually enriching interactions; to encourage interdisciplinarity in intellectual and cultural history; and to globalize the field, both in geographical scope and in subjects and methods. This series is open to work on a range of modes of intellectual inquiry, including social theory and the social sciences; the natural sciences; economic thought; literature; religion; gender and sexuality; philosophy; political and legal thought; psychology; and music and the arts. It encompasses not just North America but Africa, Asia, Eurasia, Europe, Latin America, and the Middle East. It includes both nationally focused studies and studies of intellectual and cultural exchanges between different nations and regions of the world, and encompasses research monographs, synthetic studies, edited collections, and broad works of reinterpretation. Regardless of methodology or geography, all books in the series are historical in the fundamental sense of undertaking rigorous contextual analysis.

Published by Palgrave Macmillan:

Indian Mobilities in the West, 1900-1947: Gender, Performance, Embodiment
By Shompa Lahiri

The Shelley-Byron Circle and the Idea of Europe
By Paul Stock

Culture and Hegemony in the Colonial Middle East
By Yaseen Noorani

Recovering Bishop Berkeley: Virtue and Society in the Anglo-Irish Context
By Scott Breuninger

The Reading of Russian Literature in China: A Moral Example and Manual of Practice
By Mark Gamsa

Rammohun Roy and the Making of Victorian Britain
By Lynn Zastoupil

Carl Gustav Jung: Avant-Garde Conservative
By Jay Sherry

Law and Politics in British Colonial Thought: Transpositions of Empire (forthcoming)
By Shaunnagh Dorsett and Ian Hunter, eds.

Sir John Malcolm and the Creation of British India (forthcoming)
By Jack Harrington

The American Bourgeoisie: Distinction and Identity in the Nineteenth Century (forthcoming)
By Sven Beckert and Julia Rosenbaum, eds.

Benjamin Constant and the Birth of French Liberalism (forthcoming)
By K. Steven Vincent

Character, Self, and Sociability in the Scottish Enlightenment (forthcoming)
By Thomas Ahnert and Susan Manning, eds.

Carl Gustav Jung

Carl Gustav Jung

Avant-Garde Conservative

Jay Sherry

6/24/11

Dear Helene,
You have always been a wonderful colleague. I look forward to talking about the book (I promise there will be no quizzes!)
All the best,
Jay

palgrave
macmillan

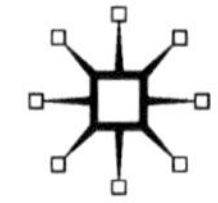

CARL GUSTAV JUNG

First published in 2010 by
PALGRAVE MACMILLAN®
in the United States—a division of St. Martin's Press LLC,
175 Fifth Avenue, New York, NY 10010.

Where this book is distributed in the UK, Europe and the rest of the world, this is by Palgrave Macmillan, a division of Macmillan Publishers Limited, registered in England, company number 785998, of Houndmills, Basingstoke, Hampshire RG21 6XS.

Palgrave Macmillan is the global academic imprint of the above companies and has companies and representatives throughout the world.

ISBN: 978–0–230–10296–5

Library of Congress Cataloging-in-Publication Data

Sherry, Jay, 1950–
Carl Gustav Jung : avant-garde conservative / Jay Sherry.
p. cm.—(Palgrave studies in cultural and intellectual history)
ISBN 978–0–230–10296–5 (alk. paper)
1. Jung, C. G. (Carl Gustav), 1875–1961. 2. Psychoanalysts—Europe—Biography. 3. Jungian psychology. 4. Europe—Politics and government.
I. Title.

BF109.J86S54 2010
150.19'54092—dc22 2010012673
[B]

A catalogue record of the book is available from the British Library.

Design by Newgen Imaging Systems (P) Ltd., Chennai, India.

First edition: October 2010

Transferred to Digital Printing in 2011

To the memory of my parents,
Frank and Sophia Sherry

Contents

Acknowledgments

If I had had to live by the dictum "publish or perish" then the words you are reading now would be coming to you from beyond the grave. Fortunately, I did not and they are not. This book was a long time in the making so there have been many people who have had a hand in its creation. They come from several different worlds each of which I want to recognize in turn.

The idea for this book germinated in Jungian soil. I want to thank Aryeh Maidenbaum and Steve Martin for inviting me to join the "Lingering Shadows" project that gave me the motivation to deepen my research. Through them I met Mike Adams with whom I had many stimulating conversations. It was around that time that the Van Waveren Foundation gave me two consecutive grants. Thank you, Olivier Bernier; as the Swiss say "Geduld bringt Rosen (patience brings roses)." I want Sonu Shamdasani to know how much I have valued his friendship and scholarly example over the years; thank you for taking time out of a very busy schedule to read the final manuscript. I also want to thank Ernst Falzeder for his hospitality, translations, and insights. John Kerr gave an early formulation of my thesis a close reading that helped redefine its focus. I want to thank the Jung family, especially Peter Jung and Ulrich Hoerni, for graciously allowing me access to their grandfather's personal library. Finally, I want to express a very special note of appreciation to Michele McKee and David Ward from the Kristine Mann Library. Their support and good humor sustained me through many years, thank you, Michele and Dave.

From academia I want to thank Geoffrey Cocks for his encouragement and reading of an early draft of the manuscript. Walter Struve generously shared his knowledge of Count Keyserling and the Weimar intellectual scene as well as advice about how to move the research along. Professor Meinardi's CUNY graduate seminar on "Romanticism Revisited" was pure pleasure and provided me with the opportunity to present my research on Jung's art work. I want to thank Lionel Gossman for sharing his deep knowledge of European cultural history and for his ongoing support. At the Freie Universität Berlin, Professors Christoph Wulf and Gunter Gebauer guided my dissertation to its successful conclusion, thank you. My colleagues at the Psychohistory Convention heard the first version of several chapters and gave me valuable feedback. I especially want Jerry Kroth to know how much I enjoyed our annual lunches where I had a chance to share my latest ideas. I also need to thank the many other authors who, although unnamed, fielded my inquiring phone calls or e-mails and graciously gave of their time and expertise.

I want to thank the dedicated staff at various libraries and collections; in Switzerland they include those at the ETH, The Central Bibliothek Zurich, the Psychology Club Zurich, and the University Library Basel. In particular, I want to thank Paul Jenkins at the Basel Mission for giving me access to Jacob Hauer's personnel file and a local perspective on his city's most famous native-son. Closer to home the staffs at the New York Academy of Medicine and the New York Public Library helped me locate sources that immeasurably enriched my understanding of the cultural context of Jung's life and work. I want to thank Deirdre Westgate for her translations from French and Hilario Barrero for his from Spanish.

Among the many friends and colleagues who have shown their support I can, unfortunately, mention only a few by name. I want to start with Brad TePaske whose belief in what I was doing served as my North Star. Jonathan Molofsky ("*l'homme extraordinaire*") processed with aplomb my mix of excitement and doubt that characterized the early stages. Renato da Silva has given more recent encouragement. Wedigo de Vivanco gave sage advice as well as bringing enthusiasm and multiple perspectives to our discussions about Germany. Florian Galler helped me think more critically about my thesis with his probing questions and Swiss perspective. Bob Kalb, Park Slope's Gutenberg, gave incredibly generous support at critical junctures throughout this project. Finally, I want to remember Tina Poremski Dalal who, once upon a time, told me that if I was into dreams I should check out "this guy Jung."

At Palgrave Macmillan Suzanne Marchand diligently read the manuscript and helped me discover the book that was embedded in it. Thank you, Sue, for everything! I want to thank Chris Chappell and Samantha Hasey who helped keep the revision process on track with her clarifications, reminders, and encouragement.

I don't know how to begin explaining what Charlie Boyd means to me and to this book. Our friendship has now spanned many years and different locales. I often imagined him as my "Deep Throat" since I never could have accessed all the information contained in this book without him. His research into the most obscure reference often hit a vein of gold. His understanding of what Jung was expressing in his native German is simply remarkable. We did not always agree on interpretations so the book's conclusions and any errors it may contain are strictly my own.

Finally, I want to thank my daughter Dana and wife Linda for putting up with this project for so long. Dana grew up with it and I will always cherish her childhood drawing of Jung's house dream. Words cannot express what I owe to Linda. All I will say here is that her technical savvy and uncommon common sense helped make this book a reality.

Introduction

Carl Gustav Jung has always been a popular thinker but never a fashionable one, his photograph more likely to adorn an art department bulletin board than one in psychology. His ground-breaking theories about dream interpretation and psychological types have generally been overshadowed by the allegations that he was, at worst, an anti-Semite and Nazi sympathizer or, at best, a mystic-philosopher uncomfortable with the hard-won truths of psychoanalysis. Most accounts have unfortunately been marred by factual errors and quotes taken out of context that have generally been due to the partisan sympathies of those who have written about him. Most biographies of Jung have taken a "Stations of the Cross" approach to his life based on the chapter topics found in his famous autobiography *Memories, Dreams, Reflections* (1989) while he usually makes only a token appearance in histories of psychology.

Things began to change with the 1970 publication of Henri Ellenberger's *The Discovery of the Unconscious*, which clearly established the Swiss context of Jung's life and convincingly located his place in the field of Swiss psychiatry. Six years later in *Freud and His Followers* Paul Roazen presented a more balanced view of the early history of psychoanalysis and more recently Sonu Shamdasani has developed the non-"Freud-o-centric" view of Jung in such works as *Jung and the Making of Modern Psychology* (2003) and in his editorial work on Jung's *Red Book* (2009).

There is, however, another, more sensationalist portrayal of Jung now abroad, one made popular by Richard Noll in his books *The Jung Cult* (1994) and *The Aryan Christ* (1997). In them Jung comes across as an occult charlatan and his institutes as Ponzi scheme training programs. Noll claims to have investigated the "German" side of Jung's life and thought, but in fact he has only done a superficial job of reconstructing this important context. To cite just one example, Noll characterized the venues for Jung's German-language publications as comparable in their intellectual depth to *Reader's Digest* when, in fact, they were publishing works by such literary figures as André Gide and Ernest Hemingway.

That Noll's lurid portrait of Jung is now the generally accepted one shows that the new Jung scholarship has not been recognized by historians who rely almost exclusively on Noll for their interpretations. They are unaware that in *Cult Fictions* (1998) Shamdasani demolished Noll's claim that Jung founded a cult. Eugene Taylor's *Shadow Culture* (1999) shows how in tune Jung was with the transcendental strain in American psychology, which explains his affinity with William James and James Jackson Putnam. Robert Brockway's *Young Carl Jung* (1996) made the first, if flawed, effort to counter the most egregious of Noll's assertions about Jung's formative intellectual experiences.

The discourse involved here concerns the master narrative of modernity, in particular how it unfolded in its psychoanalytic incarnation where Jung is invariably cast as a scoundrel; in *Putnam Camp* (2006) George Prochnik describes Jung as "simply monstrous" but "often charismatic and occasionally brilliant." In *No Place of Grace* (1981) Jackson Lears explored the antimodern sentiment of that period and although he focused on American culture much of what he said applies to its European counterpart. Modern society was becoming increasingly organized by rational, scientific methodologies that were deterministic in nature and resulted in a diminished sense of personal autonomy. Besides triggering consumerist behavior throughout all social classes, this sense of personal disorientation led to a search for new, deeper meanings that created interest in spiritualism, the occult, and non-Western art. The unconscious was being discovered as a new source of energy capable of transforming personality. When psychoanalysis is located in this cultural context Jung's ideas about the human mind seem less suspect. In *A Science for the Soul, Occultism and the Genesis of the German Modern* (2004) Corinna Treitel locates his pioneering effort to understand spiritualism psychologically as one line of development in the fin de siècle search for "alternate modernities."

I began my research wanting to know whether this member of Sgt. Pepper's Lonely Hearts Club (top row between W.C. Fields and Edgar Allan Poe) was really a Secret Architect of the Holocaust. An early source of inspiration for me was Peter Homans' *Jung in Context* (1979), which encouraged deep engagement with three core themes in Jung's thought: his interpretations of psychoanalysis, modernity, and Christianity. Modernity became my main focus as I widened the lens from the narrower issue of anti-Semitism/Nazism to a broader consideration of his evolving views on culture, politics, and race. I began to track how the opinions he first expressed in fraternity lectures in the 1890s developed through a career that lasted well into the 1950s.

My thesis is that Jung can be best understood as an example of an "avant-garde conservative" intellectual. His cultural sensibilities were decisively shaped by the neo-romantic movement dominant during his university years. In art its adherents rejected naturalism, favoring symbolism and art nouveau. In politics they deplored parliamentary democracy and rejected socialism; they preferred a Nietzschean elitism that was usually vague in concrete details. Much of my background for this came from Fritz Ringer's *The Decline of the German Mandarins* (1990), Fritz Stern's *The Politics of Cultural Despair* (1974), and Walter Struve's *Elites Against Democracy* (1973). My title evokes the contradictory ideological impulses discussed in Jeffrey Herf's *Reactionary Modernism* (1984) although this book deals more with the nexus of politics and culture and less with technology than his.

Jung's fraternity lectures and autobiography make clear how much of his worldview was formed early in his life, and the important role his reading of Eduard von Hartmann played in his intellectual development. Now best remembered for his *Philosophy of the Unconscious* (1869) Hartmann was also a critic of certain trends in German society; for him and for many others Jews were to blame for a new, materialist spirit that they saw permeating their country's economic and cultural life. They felt that national unification had led to social and spiritual fragmentation. Jung would later criticize Freud not for being Jewish per se but for contributing to the triumph of a rationalistic, "disenchanted" view of modern life. For Jung, Freud was an exemplar of the modern, Jewish intellectual who had lost his religious faith and rejected all religion as a superstitious crutch. Jung argued that this was a mistake and that a personal exploration of religious traditions was a valid path to psychological healing. Jung came to the conclusion that Freud's view of the psyche was too constricted by the positivist assumptions of nineteenth-century science and was so preoccupied with the infrared end of the human mind that he ignored the ultraviolet zone of the spectrum. Jung incurred the wrath of Freudians and the wider scientific community for his interest in parapsychology and religion.

What was it, then, that attracted Jung to psychoanalysis in the first place? As a psychiatrist he found it to be a powerful tool for understanding and helping patients. Like the bohemian intellectuals of the time he saw it as something more, however. Psychoanalysis could serve as a tool for cultural transformation and as such was one more expression of the popular life-reform movement that called for, among other things, more natural foods, more sensible clothes, and a more humane form of child-rearing. This avant-garde side of Jung's is evident in his correspondence with Freud and in the active imaginations he painted in his *Red Book*. Young talents were being encouraged to tap into the hidden sides of their own personalities, confront the emotionally charged fantasies they encountered there, and then to create art and a new society. To stimulate their imaginations they turned for inspiration to the "primitive" forms and vivid colors found in the art of Byzantium, of tribal cultures, children, and the insane. It was a time of huge public pageants and "pagan routs" where the young-at-heart frolicked. There was a feeling that Nietzsche had initiated a revolution in the fields of art, dance, philosophy, and theater. It was the age of Nijinsky and Max Reinhardt, Kandinsky and Kirchner, Steiner and Bergson. Jung's avant-garde inclinations were perhaps most prominently on display during his 1913 trip to New York where he had his portrait done by Kahlil Gibran, visited the famous Armory Show, and socialized with the Heterodoxy Club, the founding group of Greenwich Village feminists.

After his break with Freud, Jung was not content to only become the solitary sage of Zurich but actively promoted his psychology in Germany. In the 1920s he began to attend conferences at the School of Wisdom founded by Count Hermann Keyserling. There he met Prince Karl Anton Rohan and became active in the Kulturbund, a pan-European organization that sponsored cultural events and an annual conference. Jung published frequently in the Prince's *Europäische Revue*, one of the leading neoconservative journals of the day, and his work appeared in a number of other publications that shared a similar orientation. The full extent of Jung's conservative connections became clear after I studied Armin Mohler's *Die Konservative Revolution in Deutschland 1918–1932* (1994).

During this period he joined the General Medical Society for Psychotherapy, a mostly German group of doctors who practiced an eclectic set of therapeutic methods. Most of the controversy surrounding Jung involves what he said and did after becoming president of the organization in 1933, the year Hitler became chancellor of Germany. Although he never subscribed to the biological racism of the Nazis he did believe there were fundamental psychological differences between Aryans and Jews. His estimate of Nazism must be understood in light of his theory of archetypes, which he felt governed the life of nations as well as individuals. He initially interpreted the Nazi Movement as a manifestation of the "Wotan" archetype that had been reactivated in Germany. Essentially he saw it as a religious phenomenon that had to be given a chance to express itself. In this he showed his involvement in a particular trend in German church history, namely the "free church" movement that began in the late nineteenth century and was an amalgam of several small groups whose orientation ranged from socialist to pagan revivalist. After 1933 it coalesced into the German Faith Movement and sought official denominational status along with Protestantism and Catholicism. By the late 1930s Jung began to take a more critical view of Hitler but still made the callous suggestion to an American reporter that the only "cure" for Hitler was for him to invade the Soviet Union.

During World War II Jung became acquainted with Allen Dulles who was in Switzerland to coordinate espionage activities and would later become head of the CIA. Jung provided a psychological analysis of Nazi leaders and made suggestions regarding Allied propaganda. He later benefited from this relationship when he got articles published in two European journals that received secret CIA funding. Several years after the war ended he published a major article in a conservative Swiss journal that became *The Undiscovered Self*, a critique of collective psychology that was one of his last major works and that clearly reveals his credentials as an intellectual Cold Warrior.

This book will document for the first time a complete and accurate account of what Jung wrote about Jewish psychology and Nazism while placing his opinions in the wider intellectual context of the period. His writing style can be digressive but he did pursue an argument that had its own logic even though this requires following it through documents written over a period of years. This book will also call into question the work of R.F.C. Hull, the English translator of Jung's *Collected Works.* I discovered that he took many liberties with the texts and this has skewed interpretations of Jung by authors who rely on these translations. Words are changed or deleted; in at least one case a critical passage was rewritten. Most of these are to be found in volume ten, *Civilization in Transition,* which contains most of the articles relevant to the subject of this book (this volume was released in 1964, three years after Jung's death). It seems clear that Hull's translations were a deliberate effort to sanitize what Jung said in light of the controversial views that he had expressed.

For Jung, the goal of analysis was not just adjustment to life tasks but the cultivation of a symbolic attitude to help modern individuals cope with living in a "disenchanted" world; a sustained effort to understand what seems most subjective in oneself (dreams, fantasies, memories) leads to greater appreciation for the objective forces that lie beyond the individual. Jung called this process "individuation" and it takes on a special character in the lives of creative individuals such as Jackson Pollock.

As a Swiss, Jung was more comfortable with renovating than with building anew, feeling that lasting change had to be based on continuity rather than on radical experimentation. It was when Jung applied his ideas to social change that their more conservative implications became evident. When I've been asked about Jung's politics the most pithy and accurate thing I have found myself saying is that had he been born American he would have been a lifelong Republican and if British, a Tory. The public statements he made about psychoanalysis and Nazism were not just "politically incorrect" but were the most compromising expression of his long-held opinions about the role of the religious impulse in the life of the individual and of society. Both avant-garde and conservative sensibilities run through Jung's long life, the double-helix of his cultural DNA.

Chapter 1

Basel Upbringing

In 1923 Jung began to build his famous tower at Bollingen on the shore of Lake Zurich. There he could retreat from the social and professional demands associated with his home down the lake in Küsnacht and satisfy his deep need for introversion. To avoid the distractions of everyday modern life he deliberately did without such things as plumbing and the telephone. He chopped wood for his stove and used an outhouse built a short distance from the tower. Attuning himself to these simple activities and to the natural rhythms of the seasons fostered the creativity that found expression in his art work, stone carvings, and in his voluminous writings. Jung begins *Memories, Dreams, Reflections* (hereafter *MDR*) with the statement "My life is the story of the self-realization of the unconscious."[1] The tower was the realization of the first systemic fantasy that he had ever experienced and occurred when he was a boy in Basel. While walking along the Rhine on his way to school he imagined the city as situated on a huge lake from which arose a rocky hill. "On the rock stood a well-fortified castle with a tall keep, a watchtower. This was my house."[2] That it contained a library and an alchemical laboratory prefigured his activities at Bollingen where he was to carve his famous stone with alchemical inscriptions in Latin and Greek.

It is often remarked that an appreciation of Jung's Swiss heritage is necessary for a true understanding of his ideas. Unfortunately, this has rarely gone beyond the level of such clichés as Switzerland's neutral role in European affairs and its central location in continental geography. Mention is often made of Jung's explicit incorporation of a historical perspective into his theory of the psyche but this is not supported with concrete examples or a thorough exposition. What follows aims to do just that. Jung came of age in the 1890s and was deeply affected by his Basel upbringing, one that was paradoxically both parochial and cosmopolitan. Sitting astride the Rhine River the city is a

Swiss enclave with France on its western border and Germany on its eastern. Fiercely proud of such cultural traditions as its local dialect, Basel kiosks featured newspapers from its neighbors and its businessmen established offices throughout Europe and around the world. Jung's Basel upbringing was the emotional and intellectual foundation for his later achievements. The son of a Swiss Reformed minister he watched the consequences of the loss of faith in his own family and turned to philosophy to help find answers to his questions about life's meaning.

The psychology that Jung was to forge reflected his country's talent for balancing global outreach with a deep attachment to local traditions. He was proud to be from a city that had retained many of its old customs, the most famous of which was the pre-Lenten carnival known locally as "Fassnacht."

> In my native town Basel every year on January 13th, three masked dancers, a griffin, a lion, and a wild man, come down the Rhine on a raft, they land and dance around the town and no one knows why. It is an amazing thing in a modern town. These things originate before mind and consciousness. In the beginning there was action, and only afterwards did people invent opinions about them, or a dogma, an explanation for what they were doing.[3]

Jung later addressed the issue of the Swiss national identity in a review. "Does neutral Switzerland, with its backward, earthy nature, fulfil any meaningful function in the European system? I think we must answer this question affirmatively. The answer to political or cultural questions need not be only: Progress and Change, but also: Stand still! Hold fast!"[4] Jung is here echoing Jacob Burckhardt who lamented the vulgarization of culture in an age of mass democracy. Susan Hirsch notes that

> Conservatives in Basle were part of the city's intellectual and cultural elite, based as much on German immigrants and business families as on "old" aristocracy. Their allegiance was therefore more to Basle or even Germany than to Switzerland, and their brand of conservatism was truly "cultural pessimism." (Nietzsche who worked under Burckhardt and for some time held the chair of Classical Philology at Basle University, is often included in this group).[5]

He was deeply concerned about the negative consequences resulting from the break with cultural traditions that was one of the characteristics of modernization.

> We are very far from having finished completely with the Middle Ages, classical antiquity, and primitivity, as our modern psyches pretend. Nevertheless, we have plunged down a cataract of progress which sweeps us on into the future with ever wilder violence the farther it takes us from our roots... it is precisely

> the loss of connection with the past, our uprootedness, which has given rise to the "discontents" of civilization.[6]

Jung lived his life and dedicated his therapeutic praxis to connecting modern consciousness with humankind's trove of myths and symbols; he was convinced that this would promote psychologically healthy individuals and, in due course, healthier societies.

Jung built his tower in stages that he correlated with various significant personal experiences. A tower symbolizes retreat, isolation, and security.[7] To be satisfied with this image alone, however, would ignore Jung's view that symbolic dynamics spring from the tension of opposites. Polarity is a core aspect of Jung's model of psychic functioning and derived from his deep immersion in Goethe's literary and scientific writings. It later received cross-cultural confirmation when he was introduced to Taoist thought by the Sinologist Richard Wilhelm. Symbols become clichés if they do not include their opposite. In this case the tower's mythic associations need to be complemented with what "historical" meaning the tower had for Jung. Decorating it with his families' coats-of-arms, it was a place where he could commune with the spirits of his ancestors. Jung made it clear that the tower was connected with the dead (he began it shortly after his mother's death and completed it in 1955 after the death of his wife Emma).

When the annex was being built in 1927, his daughter sensed the presence of death and when the foundation was dug a skeleton was found. They were the bones of a French soldier killed in 1799 in one of the military campaigns conducted in Switzerland in the aftermath of the French Revolution. They remind us that Switzerland did not stand in splendid isolation from the major developments of European history. It had been invaded by Napoleon who had reorganized its loosely affiliated cantons into the Helvetic Republic. In the 1830s and 1840s the struggle between liberal and conservative forces created a federal system that was more firmly established by the Constitution of 1874. The first economic changes from the Industrial Revolution took place in transportation with large-scale industries developing later in Basel (chemicals) and in Zurich (machine tools). Switzerland became the transportation hub of Europe after it opened the St. Gotthard Tunnel in 1881. It established a national rail system that reached to every corner of the country and facilitated its popularity as a tourist destination. In fact, the St. Gallen-Herisau-Rapperswil railway line passes less than a hundred yards from Jung's tower and so the tranquility of the spot is regularly punctuated by the sound of passing trains.

To understand Jung, then, we must explore the influence that his Basel upbringing had upon the development of his life and thought. This is best done by looking at his family history, his formative experiences, and the intellectual currents that influenced him during his years of study at the gymnasium and university.

Jung's Family Tree

Jung was descended from two well-known Basel families. His paternal grandfather and namesake Carl Gustav Jung (1794–1864) was born in Mannheim, Germany, and studied medicine at the universities of Heidelberg and Berlin. He came under the influence of two leading liberal Protestant theologians Jacob Fries and Friedrich Schleiermacher. Besides a close personal relationship there was a social connection as well—Jung's uncle Johann Sigismund von Jung was married to Schleiermacher's younger sister. Due to Schleiermacher's influence, Jung renounced his Roman Catholic faith and converted to Protestantism. He was to suffer for his liberal views in the aftermath of the 1819 assassination of the conservative playwright August von Kotzebue by a university student. Prussian authorities cracked down, punishing many like Jung who left for Paris after spending thirteen months in prison. There he met the famed scientist Alexander von Humboldt who helped him secure a position at the University of Basel. Jung was only one of a group of émigré scholars who were hired to help restore the University's prestige after a long period of decline. Another scholar who left Berlin and joined the faculty was W.L.M. De Wette who later taught both J.J. Bachofen and Jacob Burckhardt.[8]

Schleiermacher and De Wette concerned themselves with the role of feelings in human experience. "To many theologians and to numerous pastors Schleiermacher's theology appeared to be an unnecessary concession to the pantheistic trend of German idealist philosophy. They recognized Schleiermacher's enthusiasm for Schelling, and beyond that they understood that Schleiermacher's participation in liberal politics was not unrelated to his theological outlook."[9] As his autobiography makes clear, Jung was deeply involved in religious issues throughout his life, due in great part to the fact that, like Burckhardt and Nietzsche, he came from a family of parsons. He was intimately familiar with the debates that were taking place as theologians began to face the implications of the theory of evolution for religious faith.

De Wette used a Kantian epistemology based on *Ahnung* ("presentiments") to find a place for nature and the nonrational in Christian theology. *Ahnung* refers to the intimations beyond the familiar zone of rational consciousness. Such presentiments became a preoccupation of Romantic poets, artists, and philosophers. Mesmerism, dreams, and madness were all topics of widespread interest. This trend widened when a large middle class audience was drawn to spiritualism in the late nineteenth century. All these developments had a direct influence on the theoretical goals that Jung sought to achieve in reconciling the conflicting demands of religion and science (presentiments found their way into Jung's typology of consciousness as the psychic function of "intuition"). He later wrote

> The parallelism with my psychological conceptions is sufficient justification for calling them "Romantic." A similar inquiry into their philosophical antecedents would also justify such an epithet, for every psychology that takes the psyche as "experience" is from the historical point of view both "Romantic" and "Alchemystical." Below this experimental level, however, my psychology is scientific and rationalistic, a fact I would beg the reader not to overlook.[10]

Throughout his life, Jung gave primacy to personal subjective experience over belief in the orthodoxies of either religion or science. In one of his lectures to his Zofingia fraternity he acknowledged the theologian Albrecht Ritschl's (1822–89) "extremely developed Kantian epistemology based on a solid foundation of Lutheranism."[11] Later he continued that "Ritschl rejects any illuministic or subjective knowledge, and consequently also rejects the *unio mystica*, that object on which all medieval mysticism was focused..."[12] "For almost two thousand years, from its birth in the theology of John until its decline in the philosophy of Schopenhauer, that dangerous interpretation of Christian faith which formed the foundation of the medieval world-view has fascinated the most distinguished minds."[13] At the same time, Jung made the point that the one great goal of religion was "the inner spiritualization of the individual."

Joining the Zofingia Society was an important social and intellectual milestone for Jung. It was his first real opportunity to develop close personal friendships and served as a forum in which to express he first opinions about the relationship of religion and science. The Society had been founded early in the nineteenth century and shared the patriotic-liberal philosophy popular with the German fraternities of the time. By the end of the century the Basel-City section's total of five hundred twenty-two members meant that it had provided far more members than any other canton (38 percent to the next largest 8 percent). Its membership roll was filled with family names long distinguished in the city's history: Barth (the theologian Karl was a native), Bernoulli, Burckhardt, Heusler, Iselin, Jung, Preiswerk, and Stählin. The founder of the Psychology Club of Basel Kurt von Sury had also been a fraternity brother of Jung's.[14]

In rejecting Ritschl's position Jung was adopting an attitude toward religious experience that started with Schleiermacher and De Wette. In a 1952 letter to Henry Corbin he wrote, "The vast, esoteric, and individual spirit of Schleiermacher was a part of the intellectual atmosphere of my father's family.

I never studied him, but unconsciously he was for me a *spiritus rector*."[15] Jung's affinity is made clear by his interest in the work of Rudolf Otto who is now best remembered for *The Idea of the Holy* (1917). In it Otto identified the primary religious impulse as the experience of the *numen*, a power outside oneself that engenders a feeling of awe, dread, or heightened emotion. Jung expressed just how important the term had become for him when he wrote, "The main interest of my work is not concerned with the treatment of neuroses but rather with the approach to the numinous. But the fact is that the approach to the numinous is the real therapy and as much as you attain to the numinous experiences you are released from the curse of pathology."[16] (Otto also happened to be the person who suggested to Olga Froebe-Kapteyn the name Eranos [Greek for "shared feast"] for the conference she was starting.[17])

In 1899 Otto had brought out a centennial edition of Schleiermacher's *Addresses on Religion* and four years later an essay "How Schleiermacher Rediscovered Religion." This led to his involvement in the neo-Friesian school movement initiated by Leonard Nelson, his former colleague at the University of Göttingen. Jakob Fries was another theologian of the Romantic period concerned with the place of feeling in religion. In his *Kantisch-Friessche Religionsphilosophie* (1909) Otto tried to correct the defects in Fries' decidedly idealistic system but "came to see that Fries, while presenting effectively the rational and moral foundation of religion, had missed the uniquely religious element therein."[18] This led him adopt a "history-of-religions" approach that went beyond Christianity to include an appreciation of the other religious traditions he encountered on a trip to North Africa, India, and Japan in 1911–1912.

Otto's first publication in 1898 had been on Luther's views of the Holy Spirit. The date and topic coincide with Jung's Zofingia lecture on "Thoughts on the Interpretation of Christianity, with Reference to the Theory of Albrecht Ritschl." Jung began by identifying himself as a student of medicine making a foray into theological speculation and sought to win his audience over with his erudition and blunt opinions. At issue was the controversy over whether Christ was a human or divine being. Ever since higher criticism had subjected biblical scriptures to the same scrutiny as other texts, doubts had multiplied about the authenticity of Christ's miracles and the meaning of his divinity. Lionel Gossman states that earlier in the century

> De Wette had interpreted the entire Old Testament canon, together with a good part of the New Testament, as myth rather than factual history but had sought at the same time to rebuild what he might seem to have destroyed by rehabilitating myth and symbol, in the spirit of his friend Georg Friedrich Creuzer, the widely read Heidelberg philologist and mythologist, as valid sources of historical understanding.[19]

Although he rejected traditional explanations of biblical events De Wette still accepted Christ's divinity. The radical theologian David Strauss (1808–1874) took the next step in his *Life of Christ* (1835).

> Strauss took sides with those who denied all historicity to the supernatural events of Jesus' life as described in the Gospels. Not that Jesus himself had not lived and died; Strauss conceded that Jesus was a real person. But not only had supernatural events had not occurred, there was no point in seeking a specific natural or historical event of any kind behind New Testament events. They were, in a real sense, fictional events.[20]

This approach was later continued by such "demythologizers" as Albert Schweitzer whose *Quest for the Historical Jesus* appeared in 1906. There he wrote that "few understood what Strauss's real meaning was. The general impression was that he entirely dissolved the life of Jesus into myth."[21] This was the background to Jung's lecture in which he concluded that "Christ is a metaphysical figure with whom we are bound in a mystical union which raises us out of the sensory world."[22] Here he is articulating in philosophical language a concern for the symbolic that would define his career as a medical psychologist.

In opposing Ritschl's rationalistic position, Jung relied on the philosopher Eduard von Hartmann (1842–1906) whose bestseller *Philosophy of the Unconscious* had incorporated Schopenhauer's Will into the idealistic tradition. Jung wrote:

> I call on everyone, and especially theologians, to remember the truth that Eduard von Hartmann hurled down at the feet of all Christians, and I implore that they harken to his voice: "The world of metaphysical ideas must always remain the living fountain of feeling in religious worship, which rouses the will to ethical action."[23]

Citing von Hartmann a little earlier he said, "What is so special about Christ, that he should be the motivational force? Why not another model—Paul or Buddha or Confucius or Zoroaster?...If we can view Christ as a human being, then it makes absolutely no sense to regard him as, in any way, a compelling model for our actions?"[24] Jung then discussed the nature of Christ's divinity not in terms of Christian dogma but in terms of his emotional impact on his early circle and millions since focusing on the subjective experience of Christian believers.[25] Such daring speculations would have given any of his ecclesiastical relatives who came over for Sunday dinner cause for alarm.

Von Hartmann was clearly influential for Jung from the time of the Zofingia lectures until he began to revise his model of the psyche after his encounter with alchemy and modern physics in the early 1930s. In *MDR* Jung recalled that he read von Hartmann "assiduously" during his university years and later acknowledged it as a philosophical antecedent of his psychological theory of the unconscious. Jung's personal library contains four of the philosopher's books: *Philosophy of the Unconscious* (1872), *German Aesthetics since Kant* (1886), *Modern Psychology* (1901), and *The World-View of Modern Physics* (1909).

Von Hartmann's influence on Jung's post-Kantian epistemology has been recognized but not his deep and ultimately problematic influence on Jung's ideas about religion and society.[26] Like many German intellectuals of his time, von Hartmann became decidedly conservative-nationalist after the founding of the German Reich in 1871. He saw the spread of materialism, democracy, and socialism as threats to Germany's cultural identity. He shared this uneasiness about Germany's modernization with another cultural pessimist Paul de Lagarde (1827–1891) who was Ritschl's main antagonist on the Göttingen faculty, criticizing him for, among other things, his rationalistic dismissal of mysticism.

Von Hartmann, Paul Means argues,

> spoke of future religion as a new creation of the Indo-Aryan spirit expressing a pan-tragic sense of life. His disciple, Arthur Drews, who accepted Hartmann's "Philosophy of the Unconscious" as a new revelation, in 1910 surprised the German theological world by his book *Die Christusmythe*' which argued the non-historicity of Jesus, and sought the foundations of the Christ-cult largely in Aryan mythology and legend.[27]

Jung was to closely follow the work of Drews, von Hartmann's most devoted disciple, and was later friendly with *his* student Leopold Ziegler. Jung echoed some of the views of this school of thought in his lecture when he mused about "the disgrace of a Germany overcome by materialism"[28] and insisted that "The Germanic variety of the species *Homo sapiens* has a reputation for sensibility and depth of feeling."[29]

Jung was not only, or even most importantly, a Germanic nationalist. His views reflect those held by many in the Protestant mandarin class of fin de siècle Switzerland and Germany. The development of Social Democratic parties representing the interests of the working class created anxieties about the future course of social relations. He lambasted proponents of a strictly materialistic philosophy of science "for having stuffed a passel of materialistic rubbish into the gaping mouths of those guttersnipes, the educated proletariat."[30] This snobby diatribe would have found favor with many of his fraternity brothers who came from the same social background as Jung. They were all familiar with the perspective of Jacob Burckhardt who lamented the vulgarization of culture in the age of mass democracy.

In Jung's summer 1898 lecture to the Zofingia there is the thoroughly Burckhardtian passage, which is as follows:

> Modern man knows nothing of the individual. The individuals he knows are cantons and nation-states. As a rule he has already lost his consciousness of himself as an individual. He feels that he is an atom, a mere link in an endless chain that makes up the state. Modern man shifts responsibility for the creation of individual happiness from himself to the state... modern man seeks to level, that is, wipe out, individuality by educating everyone, as much as possible, to be exactly the same.[31]

For Burckhardt, Jung, and other Baselers, the state was not some abstract concept but the Swiss federal government that had periodically expanded its powers at the expense of the cantons. For Basel the most traumatic instance of this occurred in 1833 when federal authorities intervened to end the three-year conflict between Basel and its outlying rural districts. The federal government supervised the creation of two separate cantons Basel-Stadt and Basel-Land in a settlement that involved a division of financial resources and obligations. Although this resolution insured the continued influence of the old ruling families, it was widely resented in Basel as undue meddling in a strictly cantonal affair. Jung was to remain loyal to the political values of his native city throughout his life, asserting in a 1936 interview that "A decent oligarchy—call it an aristocracy if you like—is the most ideal form of government."[32]

The Preiswerks and Basel's Religious Milieu

Jung's Preiswerk relatives belonged to the orthodox wing of the religious party in Basel (eight of his uncles were pastors). "My uncle and cousins... seemed safely ensconced in a self-evident world-order, in which the name of Nietzsche did not occur at all and Jacob Burckhardt was paid only a grudging compliment. Burckhardt was 'liberal,' 'rather too much of a free-thinker'..."[33] His grandfather Samuel Preiswerk (1799–1871) was pastor of St. Leonhard's Church and a Hebraist who edited a monthly journal *The Orient* in which he advocated the restoration of Palestine to the Jews. Jung's grandmother Augusta Preiswerk neé Faber (1805–1865) was Samuel's second wife and the daughter of a clergyman from Württemberg. In his doctoral dissertation about the mediumship of his cousin Helene, Jung gave a candid clinical snapshot of this branch of the family.

> The paternal grandfather was very intelligent, a clergyman who frequently had waking hallucinations... A brother of her grandfather was feeble-minded, an

> eccentric who also saw visions. One of his sisters was also a peculiar, odd character. The paternal grandmother, after a feverish illness in her twentieth year—typhoid fever?—had a trance lasting three days, from which she did not begin to awake until the crown of her head was burnt with a red-hot iron. Later on, when emotionally excited, she had fainting-fits; these were nearly always followed by a brief somnambulism during which she uttered prophecies. The father too was an odd, original personality with bizarre ideas. Two of his brothers were the same. All three had waking hallucinations. (Second sight, premonitions, etc,) A third brother was also eccentric and odd, talented but one-sided.[34]

Before considering the impact of the Preiswerk family's involvement in spiritualism on Jung, another aspect of the Basel church scene needs to be mentioned. That was the active role that Pietism played in the religious life of the city. This religious movement had developed in the eighteenth century as a reaction to the rationalism characteristic of the Enlightenment. Among its most prominent groups was the Moravian Church rooted in the Hussite movement of the sixteenth century but revitalized by the Baron Zinzendorf at his estate in Saxony in 1722. After his visits to Basel in 1740 and 1758, the Pietist movement gained a following among artisans and some clergymen. They proved to be a counterweight to the influence to the orthodox majority.[35]

This movement was found well beyond the city limits of Basel itself, extending into the adjacent states of southern Germany, especially Württemberg. Although concerned about the individual's personal relationship with God it was active in evangelical outreach, being responsible for the founding and growth of the Basel Mission, which sent missionaries to Africa, the Middle East, and India. Over the years, a significant number of the missionaries were from Württemberg (among them the parents of Hermann Hesse who was born in Calw and Jacob Hauer who taught at the university at Tübingen and was close to Jung during the 1930s).[36] This region also was the home of Justus Kerner (1786–1862) who recorded his treatment of Friedericke Hauffe (1801–1829), popularly known as the Seeress of Prevorst, in a celebrated publication that grabbed the interest of many intellectuals of the time, among them David Strauss and later young Carl Jung.[37]

Jung sprinkled his Zofingia lectures with references to the religious milieu of his home town. "In Basel there are hundreds, perhaps thousands of people with adamant faith in the miracles of the Old and New Testaments, but who would not for anything in the world admit that identical or similar events are still taking place today."[38] At a certain point the distinction between pietistic witnessing and spiritualistic manifestations blurred as séances became a popular way of contacting departed souls. Jung avidly participated in séances with his mother and cousin Helene the medium. As *MDR* makes clear, it was to his Preiswerk inheritance that Jung attributed his lifelong interest in parapsychology.[39]

Parents, Sister, and Boyhood

Jung wrote about his parents at great length in his autobiography, creating portraits that reflected his experiences of their personalities and marriage.

> As a country parson [my father] lapsed into a sort of sentimental idealism and into reminiscences of his golden student days, continued to smoke a long student pipe, and discovered that his marriage was not all he imagined it to be. He did a great deal of good—far too much—and as a result was usually irritable. Both parents made great efforts to live devout lives, with the result that there were angry scenes between them only too frequently.[40]

The pages are filled with Jung's memories of his father's religious crisis. Johann Paul Achilles was Carl Gustav's youngest son and studied philology at Göttingen where Albert Ritschl and Paul de Lagarde were on the faculty. He wrote his doctoral dissertation on the Arabic version of the Song of Songs but there are no records of his having received his degree. This might explain his decision to join the ministry and the reason why he was assigned to a series of minor rural posts rather than one in Basel, which one would expect for a person with his family connections. The first was at Kesswil on the shores of Lake Constance, the second at Laufen near the Rhine Falls, and his final post was in Klein-Hüningen a village of farmers and fishermen across the river from Basel that later came to be dominated by the nearby harbor facilities that served the many ships that plied the Rhine trade.

Jung's entry into the Gymnasium at the age of eleven created a shock. There he rubbed shoulders with the sons of patrician families and so become acutely aware of his family's circumscribed financial situation. This gave him some measure of understanding of his father's struggles and acted as a spur to his own academic achievement. He came to realize that he would have to rely on his own intellect to get ahead in the world.

Jung's development was affected by his father's inability to meet his need for explanations for deep religious questions. His father comes across as a decent man, a probable agnostic who urged his son to think *less* and *believe* more about such things. "[My father] had failed to experience the will of God, had opposed it for the best reasons and out of the deepest faith... he did not know the immediate living God."[41] Jung observed the effect of Nietzsche's proclamation "God is dead!" in his father's life and said that this insight "paved the way for modes of adaptation to my father's religious collapse as well as to the shattering revelation of the world as we see it today."[42]

Paul Jung died from cancer in 1896 at the age of fifty-four soon after his son began at the university. This led to an even more precarious financial situation for the family. His mother received a modest pension and had to move with Carl and his sister Gertrud to the old Bottminger mill on the other side of the city. To pay his tuition, Jung depended on a grant and help from one of his uncles.

Gertrud was born in 1884 when Jung was nine; her arrival seems to have caught Jung by surprise since he hadn't noticed anything unusual about his mother frequent takings to her bed. It was apparently an unwanted pregnancy since Jung remarked that "Subsequent odd reactions on the part of my mother confirmed my suspicions that something regrettable was connected with this birth."[43] Fondly remembered by her nieces and nephew Gertrud lived a quiet life first with her mother and later with her brother and his family until her death in 1935.

Jung's university years were dominated by his interest in spiritualistic phenomena, the best-known example being his attendance at his cousin's séances. Two events occurred at the Mill in the summer of 1898 that gave Jung the first-hand experience of the uncanny that he craved. One involved a round walnut table inherited from the Preiswerk family, the top of which split against the grain from its edge to past its center. The loud cracking brought people running. The other involved a bread knife sitting in a basket in the sideboard that apparently exploded into four pieces. A cutler examined the knife the day after the incident and told Jung that it had to have been deliberately broken. Jung refused to accept this explanation and for the rest of his life attributed the two occurrences to his cousin's mediumistic influence.[44]

What if the cutler was right? A strong case for this can be made from the physical evidence and from the conflicting accounts Jung gave of the incident.[45] The better-known one is that given in *MDR*, the other was in a 1934 letter to J.B. Rhine, pioneering parapsychologist at the Duke University laboratory established by William McDougall. In *MDR* Jung remembered arriving home shortly after the incident to find his mother, sister, and the maid in a state of agitation. They directed him to the sideboard where he found the broken knife. In the letter Jung said that the explosion occurred in the presence of his mother while he was in the garden, the maid was in the kitchen, and his sister was out. Although it was written many years after the letter, there are good grounds to believe that the *MDR* version more accurately reports what happened that day. We cannot accept the argument that the discrepancies are minor and merely the result of mistaken memory. One would think that such an experience of the numinous would have left an indelible memory in Jung. It seems, though, that he altered the story to bolster its credibility to another psychologist interested in the paranormal. Jung is himself present rather than having to learn about it second-hand.

In any case, the physical evidence does support the cutler's naturalistic explanation. A photo of the knife that Jung included in his letter to Rhine shows it broken in three places at even intervals suggesting that it had been deliberately done. Jung found the pieces lying in each corner of the rectangular basket and noted in the letter that there were no cuts on either the bread or the sides of the basket. This is further proof that someone had broken it and then neatly arranged the pieces.

The question now becomes *who* would have done it? The likeliest suspect is Gertrud whose presence is deleted from the Rhine version, the "missing fourth" as it were. We should try to imagine her situation. For several weeks she would have been listening to the excited discussions about the cracked table, something she had unfortunately missed out on because she'd been at school. A shy teenager, Trudi was impressed with her brother's burgeoning talents and participated in the family séances that began in 1895 but were suspended when Helly had to prepare for her confirmation. At one point Trudi fell into a trance and spoke to her brother in their father's voice.[46] One can picture her deliberately breaking the knife, timing it so that her brother could arrive home to discover the "exploded" knife. Why would she do such a thing? Although it is possible to agree with the cutler that it was intended as a practical joke, the more likely reason is that she did it as a way to figure in a family drama starred in by her more theatrical cousin. It would then be a case of deception that, unlike those of Helene that were exposed later, Jung failed to detect. But in the end, importantly, his preference for an "occult" explanation took precedence over what he considered an unduly narrow, naturalistic explanation. These experiences bolstered his confidence in challenging what he felt were the limitations of science.

Jung's mother Emilie was a strange, brooding presence in Jung's life. In two photos separated by decades she sits in an identical pose: looking out at the camera with a steady gaze, her right arm folded over her left.[47] Raised in a house where her father conversed with the spirit of his deceased first wife, she would later participate in séances with her niece that included extensive communications from that now-deceased man. She was a demanding woman unhappy in her marriage who invested her affections and ambitions in her son. Jung would later reminisce that "It was plain that she was telling me everything that she could not say to my father, for she early on made me her confidant and confided her troubles to me."[48]

Jung clearly felt that his mother was responsible for the daimonic element in his personality.

> By day she was a loving mother, but at night she seemed uncanny. Then she was like one of those seers who is at the same time a strange animal, like a priestess in a bear's cave. Archaic and ruthless; ruthless as truth and nature. At such moments she was the embodiment of what I have called the "natural mind." I too have this archaic nature...[49]

Jung identified his mother with a darkly vital pagan heritage that stood in marked contrast to his father's anemic Christianity.[50] It was his mother who encouraged him to read Goethe's *Faust*, an experience that came as a literary confirmation of his intuitive sense of the Number 2 Personality he felt he shared with her. Her voice figured prominently in the first dream he could remember, one that occurred when he was between the ages of three and four (at a time when she was away at a hospital in Basel and he suffered from a case of eczema). In *MDR* he discussed this anxiety dream of a ritual phallus in an underground chamber in great detail.[51] He related that as he gazed on the phallus he heard his mother's voice calling out from above "Yes, just look at him. That is the man-eater!"

He linked this dream phallus to the dark Lord Jesus and to the Jesuits, associations that stemmed from other formative experiences. The association to be highlighted here is that of "the Jesuits." Jung's early religious preoccupations not only shaped his own spiritual development but reflected issues that were of pressing political importance to the Switzerland of his boyhood. The first trauma of which Jung was conscious involved his encounter with a Catholic priest on the road in front of the parsonage at Laufen. One summer day he looked up to see a man coming out of the woods wearing a broad hat and a black dress.

> At the sight of him I was overcome with fear, which rapidly grew into deadly terror as the frightful recognition shot through my mind: "That is a Jesuit." Shortly before I had overheard a conversation between my father and a visiting colleague concerning the nefarious activities of the Jesuits. From the half-irritated, half-fearful tone of my father's remarks I gathered that "Jesuits" meant something especially dangerous, even for my father.[52]

The fact that the word was capable of provoking terror in such a young child suggests that it functioned as a cultural "complex-indicator" for nineteenth-century Swiss Protestants. The country's complicated denominational situation dated back to the Reformation. The rural inner cantons remained loyal to Rome while the urban middle classes opted for reform. Geneva followed John Calvin while the German-speaking towns banded together under the leadership of Ulrich Zwingli who was killed by Catholic forces at the Battle of Kappel (1531). It was at this time that Basel joined the Swiss Confederation (1501) and was reformed by Johannes Oecolampadius (1529). Although it is a common mistake to think that Jung had a Calvinist upbringing, he was raised in the Swiss Reformed Church.

The Catholic Counter-Reformation battled the Protestant Movement with the force of intellect as well as of arms. Its most effective tool was the Society of Jesus founded by Ignatius of Loyola and approved by the pope in 1540. Jesuits were active in the establishment of schools throughout Europe and its New

World and Far Eastern missions. Sent to Switzerland by the Archbishop of Milan Charles Borromeo, they founded many colleges there, among them were those at Lucerne (1574), Fribourg (1582), Bellinzona (1646), and Solothurn (1668). They became confessor-advisors to many Catholic monarchs but elicited suspicion and hostility from powerful groups within the Catholic Church. These forces were able to have the Society suppressed by Pope Clement XIV in 1773. The only community not legally bound by the decree was that in the part of Poland acquired by Russia during its partitions.

The Society was restored by Pope Pius VII in 1814 as part of the settlements made at the Congress of Vienna and in 1844 Jesuits were recalled to Lucerne. They did not stay long because they were blamed for fomenting the unrest that resulted in the Sonderbund War in 1847. Under the terms of the agreement of 1848 they were expelled from Switzerland, a policy that was later codified in the country's 1874 Constitution. Their "nefarious activities" were still the subject of emotional discussions among Protestant clergy, a conversation that little Carl overheard. The events in Switzerland were closely related to those taking place in Germany as part of Bismarck's *Kulturkampf* against the Catholic Church. Aiming to curb the powerful Catholic Center Party based in Bavaria, he carried through a series of anti-Catholic measures that included the expulsion of the Jesuits in 1872.

Jung went on to become a pioneer in the ecumenical approach to the psychology of religion. He was proud of his sensitivity to the Catholic cult of the saints, the ritual of the Mass, and the dogma of the Assumption. Still something of his old feelings seemed to play a role in his reluctance to visit Rome. He never visited there and fainted while buying a train ticket to it in 1949. He explained it as a reaction to the overwhelming effect of its classical heritage when he wrote "if you are affected to the depths of your being at every step by the spirit that broods there, if a remnant of a wall here and a column there gaze upon you with a face instantly recognized, then it becomes another matter entirely."[53] One wonders if an old, deep-seated antipathy to "Jesuitical" Rome as well as possible ambivalent feelings about his grandfather's conversion played a role in his inability to visit the city.

Between Neohumanism and Natural Science: Jung's Unique Education

Jung's formal education began under the tutelage of his father who began to teach him Latin when he was six. Like students throughout the German-speaking countries, he was grounded in the classical languages. (Ernest Jones remembered being impressed with Freud's and Jung's abilities to recite lengthy passages of Greek and Latin authors in the original.[54]) Something of its importance can be gleaned from the fact that it was tackling Latin grammar that helped Jung overcome a bout of academic indolence that occurred when he was twelve. Later, he would incorporate classical terms into his psychological vocabulary among which were the "inferior (lower)" function and the "persona." This classical tradition was rooted in the humanist curriculum established by such Renaissance thinkers as Erasmus who was a long-time resident of Basel. While there he brought out critical editions of St. Jerome, Seneca, and Plutarch followed by Greek and Latin translations of the New Testament. Maintaining a tolerant attitude amid the growing sectarian violence, Erasmus remained loyal to the Church and was forced to leave Basel in 1529.

This intellectual tradition was continued by a group of reformers who inaugurated important changes in German education in the early nineteenth century.

> The neohumanist movement in Germany, which the Basel authorities tried to import wholesale into the city at the Restoration, can be viewed as a peculiarly German version of the Enlightenment and, by the time it was being promoted by Humboldt, as a rejection of the French road to individual emancipation and social transformation by way of political revolution, in favor of an indigenous German road.[55]

Rather than a specialized technical training, neohumanism's goal was the cultivation (*Bildung*) of the individual's various talents. It formed the foundation for Burckhardt's ideal of the Renaissance Man and was one source of Jung's concept of "individuation."

Jung was an indifferent student in most school subjects and only got consistently good grades in classical languages, the study of which would have included lessons in history and expository writing. His intellectual curiosity led him to philosophy. "Kant, Schopenhauer, C.G. Carus, and Eduard von Hartmann 'had provided him with the tools of thought.' He had read their works when young, perhaps as early as his sixteenth year, at any rate well before the beginning of his medical studies, and they influenced his thinking decisively."[56] He first dreamed of being an archeologist but had to abandon this goal since there was no possibility of studying this subject at the University of Basel where courses of study were still grouped along the medieval divisions of law, philosophy, medicine, and theology.

Jung opted to study science and so entered the university's Faculty of Medicine in the spring of 1895.[57] For a boy growing up in a rural village the countryside was just a short stroll away. Jung later remembered that during his extended absence from school at the age of twelve "Nature seemed to me full of wonders, and I wanted to steep myself in them. Every stone, every plant, every single thing seemed alive and indescribably marvelous.

I immersed myself in nature, crawled, as it were, into the very essence of nature and away from the human world." This interest soon led him to subscribe to a scientific periodical and to start a collection of mineral specimens, insects, and human and mammoth bones. Jung belonged to a generation of young men who came of age with a passionate interest in unlocking the secrets of the book of nature.

It is important to understand that Jung, like many other scientists in the German-speaking world, was schooled in a tradition rooted in the scientific works of Goethe rather than in Darwin's *Origin of Species*. Goethe made several important contributions to scientific knowledge, the most famous being his discovery of the intermaxillary bone in the human skull. He was a keen empiricist but opposed to the mechanistic model proposed by Bacon and employed by Newton (whose theory of color he also opposed). He rejected a mathematically abstract approach to science for one that included both the sensual reality of the thing observed and the imaginative faculty of the observer. This technique of *Anschauung* ("direct vision") reflected Goethe's artistic-poetic temperament and was used to study Nature in a holistic, organic way.[58] A special moment in this regard came to Goethe while he was visiting the Botanical Gardens in Palermo. There amidst the exotic vegetation the concept of the *Ur-pflanze* ("primordial plant") occurred to him and provided the key to understanding the common template for explaining life's diversity.

Late in his life, Jung reminisced about the influence of this scientific approach. "My life work in historical comparative psychology is like palaeontology. That is the study of the archetypes of the animals, and this is the study of the archetypes of the soul. The *Eohippus* is the archetype of the modern horse, the archetypes are like fossil animals."[59] He wrote, "What fascinated me most of all was the morphological point of view in the broadest sense."[60] Goethe's study of the structural components of various families of living things developed into the field of comparative morphology. It was developed by a group of German scientists who rejected both strictly materialistic explanations and the overly idealistic speculations of *Naturphilosophie*. Inspired by Kant's defense of teleology, men like J.F. Blumenbach, Karl Ernst von Baer, and Rudolph Leuckart made many contributions to embryology and zoology but their "teleo-mechanistic" morphology was to be dismissed by Du Bois-Reymond as too metaphysical. It is no accident that the positivistic Du Bois-Reymond was the main scientific target of Jung's Zofingia lectures, which he gave after becoming a junior assistant to Friedrich Zschokke who had studied parasitology with Leuckart at Leipzig.[61]

Under Zschokke Jung was trained in the evolutionary theory and comparative anatomy of the teleo-mechanisitic school. He found the study of physiology repugnant because of its dependence on vivisection. Because of his "sympathy for all creatures" he found the practice cruel and unnecessary. He avoided laboratory demonstrations as often as he could and justified his decision with the thoroughly Goethean rationale that "I had imagination enough to picture the demonstrated procedures from a mere description of them."[62]

With collegiate bombast Jung heaped scorn on Du Bois-Reymond for his shallow philosophy and pernicious influence. He chided "educated people" (by indirection his fraternity-brother audience) for parroting the materialist dogmas of that "Papa" from Berlin and so demonstrating their intellectual poverty. Emil Heinrich Du Bois-Reymond (1818–1896) was a leader in the movement to reduce physiology to chemistry and applied physics. Besides his influence in the realm of university appointments, his popular writings about science reached a wide audience.[63] Elsewhere Jung wrote that Du Bois-Reymond was "A professor drowned in mechanistic psychology and nerve-and-muscle physics is sowing the poisonous seed that fecundates confused minds... Gradually the mud is seeping down from the heights of the university. The natural consequence is the moral instability of the upper echelons of society and the total brutalization of the working man."[64]

Anti-Materialism and Anti-Semitism at the Fin de Siècle

In this lecture Jung appealed directly to the social prejudices of his audience, proud sons of the Basel patrician class. He hoped to win converts to his anti-materialistic position by clearly linking materialism to one of the period's major developments, the expansion of an urban working class that sought its identity in the writings of Karl Marx and formed socialist parties to promote its interests. He was alerting his fraternity brothers to the threats to their traditional leadership status posed by different class and ideological interests.

His comments reflect the mood of cultural pessimism being expressed by some of the contemporary German writers that Jung was reading. Mention has already been made of the philosopher Eduard von Hartmann; another was the Leipzig physicist Johann Zöllner who became for Jung a martyr to science.

> In 1877 the noble Zöllner published his scientific tracts in Germany and fought for the spiritualist cause... [but] Mortally wounded in his struggle against the Judaization of science and society, this high-minded man died in 1882, broken in body and spirit... the spiteful Du Bois-Reymond defamed this cause throughout a Germany in moral decline. All in vain—the Berlin Jew came out on top.[65]

First linking "materialism" to the proletariat, Jung now identified the final element in this network of associated threats, "the Jew."

Jung owned Zöllner's *Transcendental Physics* (1879) and in the scientist's reports of his experiments with the American medium Henry Slade he found inspiration for his doctoral research.[66] Jung's choice of Zöllner as his champion for the spiritualist cause proved to be problematic, however. Besides his advocacy of spiritualism, Zöllner was an out-spoken anti-Semite, being the only professor in Germany to sign the anti-Semitic petition sponsored by Nietzsche's brother-in-law Bernhard Förster in 1880. Dubbing the nineteenth century "the century of Jewish liberalism," he deplored Jewish emancipation for injecting a foreign influence into German culture and expressed his concerns about the "reigning Judaization of German universities." That Jung adopted this line is evident from his comment about the "Judaization of science" and the "Berlin Jew." Jung's diatribe would have triggered an immediate association in his listeners who were at the moment witnessing many Berlin Jews arriving in their city to attend the First Zionist Congress.[67]

Although Jung's anti-Semitic rhetoric was borrowed from Zöllner, his opinions about Jews also reflected those held by Burckhardt and many others in Basel.[68] For many Germanic intellectuals Berlin had become the symbol of all that they found objectionable about modern life: a metropolis inhabited by a rootless population pursuing wealth and amusement. Both men criticized Berlin University for its devotion to academic specialization but from differing points of view. To Zöllner it represented a sell-out to Jewish rationalism. Burckhardt, who declined to accept a chair of history there, had reservations about the abandonment of the *Bildung* ideal and the dangers of mere technical achievement being put at the disposal of a modern nation-state. Berlin was clearly associated with "the Jew" who was singled out as the prime catalyst of the process of modernity. The uneasiness these men felt about modern society found expression in their anti-Semitic prejudices.

As a young man Burckhardt had studied in Germany where he shared the liberal enthusiasms of his friends of the generation of 1848. The lasting effect of their defeat was for him to retreat to his hometown where he dedicated himself to serving its long tradition of conservative, humanist education. The lesson he learned was to mistrust the mass forces unleashed in European society by the French Revolution. The unchecked industrial expansion created an expanding middle class whose cultural pretensions he decried. Among the philistines that he caricatured were the Jews of Frankfurt.

> People who are incapable of producing something beautiful are unable to do so whatever the style, and all the "motives" and "themes" in the world won't help

> a man without imagination. Most of what is built in Italian Renaissance style is hideous, despite its richness...And you should see the classical buildings! "For the wealthy Jew/Only caryatids will do."[69]

It is unfortunate that Burckhardt saw Jews as a threat to his cherished "old culture of Europe" since in pursuing the educational opportunities opened to them after emancipation they became its most ardent supporters. The real threat lay elsewhere. Count Harry Kessler recalled a conversation that illuminated this situation.

> 1831 saw the beginning in Switzerland of a cultural leveling process which lasted until about 1875 and produced to a varying degree detestation, fear, hate, and contempt in these men [including Burckhardt]. The lower middle class, which mistook its semi-education for culture, came to power and pushed the old, highly cultivated patrician families aside. Switzerland thus forestalled developments all over Europe...Since then Switzerland has become conservative.[70]

Basel's Iconoclasts: Burckhardt and Nietzsche

Jung's Basel upbringing left an indelible mark on him, shaping the political and cultural views he held throughout his life. Of particular importance was the influence of Jacob Burckhardt who was a one-man cultural institution and who eventually became "a kind of patron saint of Basel."[71] As Ira Progoff says

> Jung absorbed Burckhardt's historical orientation, then, not because he was a close student of society during his early years, but because Burckhardt's work and insights were part of the cultural atmosphere. Without consciously taking over any specific doctrines, the historical way of thinking about all human phenomena became part of his underlying outlook, and later on it was a natural step for him to apply an historical point of view to the analysis of psychic phenomena. In this sense, Jung's work must be interpreted as being related to the great Burckhardt tradition...[72]

In 1898 Jacob Oeri, Jung's old Latin teacher and Burckhardt's nephew, began to bring out his uncle's *Greek Cultural History.* During this time his son Albert was a fraternity brother of Jung's and involved in helping his father prepare other manuscripts that were to appear in 1905 as *World Historical Reflections.*[73] Besides seeing Burckhardt on the streets when he was a boy, Jung had first-hand access to what was in his manuscripts and later adopted several of Burckhardt's formulation in his work, most notably that about the "Archimedean point outside of events."[74]

Nietzsche's tenure (1869–1879) as a professor at the university also created a legacy in the town. His public lectures on Greek culture attracted the attention of Bachofen and Burckhardt. The latter was to recoil at the Dionysian import of the German's philosophy but maintained a correspondence with him. Jung wrote that "After this book [*Zarathustra*], they said he was mad. Jakob Burckhardt got a chill when he touched it; he squirmed away from the awful thing."[75] Burckhardt's role as a benevolent father-figure is most poignantly captured in a postcard Nietzsche wrote to him at the time of his breakdown in January 1889. "In the end I would much rather be a Basel professor than God; but I have not dared push my private egoism so far as to desist for its sake from the creation of the world."[76] It was another old Basel colleague, Franz Overbeck, professor of theology, who went to Turin to bring Nietzsche back to Basel before his return to the care of his mother and sister in Germany.

Jung's cultural views were influenced by these two thinkers ("Nietzsche's mind was one of the first spiritual influences I experienced. It was all brand new then, and it was the closest thing to me."[77]). These connections involved social as well as intellectual affinities: just as Oeri was Jung's source for information about Burckhardt, Jung heard many stories about Nietzsche from his other close friend Andreas Vischer whose parents had befriended the philosopher. Vischer was to die relatively young after serving as a missionary doctor in the Middle East. In the German edition of *MDR* Jung reminisced about a four-day trip on Lake Zurich he made in 1913 with his two old friends. It was a magical moment on board his boat as Oeri read aloud from the Nekyia ("Voyage to the Underworld") chapter of *The Odyssey*. He remembered the two as his "boon companions" whom he was lucky to have found in his youth.[78]

Jung's *Zarathustra Seminar* is laced with anecdotes acquired from them and other old Basel acquaintances.

> I knew a man whom Nietzsche considered one of his great friends. He [Friedrich von Müller] was a professor of internal medicine, a highly educated man, very musical, and Nietzsche would often go to his house—one never knew exactly when; he would appear suddenly and sit down at the piano and play for hours on end. He spoke to nobody and nobody could speak a word to him. And then he went away and said what a nice evening it had been.[79]

Such stories, usually told with an unflattering slant, circulated around Basel in the 1890s. Highlighting Nietzsche's eccentricities, they were meant somehow to explain his madness. He lived on until 1900 at Weimar where his sister Elisabeth Förster-Nietzsche set up the Nietzsche-Archive to control his literary estate and reputation, aligning it with extreme-right circles in Germany. In his doctoral dissertation Jung mentions corresponding with her to confirm a case of cryptomnesia that he found in *Zarathustra*.[80]

Jung had heard gossip about Nietzsche since he was a boy but waited until he had first read Goethe and Schopenhauer to read *Zarathustra* and found it "morbid. Was my No. 2 [personality] also morbid? This possibility filled me with a terror which for a long time I refused to admit..."[81] Jung illustrated this theme of morbidity by recalling that the only people he knew who were openly declared Nietzschean adherents were both homosexuals, one of whom committed suicide and the other went to seed as a misunderstood genius.[82] This reminiscence was contradicted by his fraternity brother Gustav Steiner in an article he wrote after *MDR* was published.

> There were a considerable number of adepts of Nietzsche... We were moved by the tragedy of the genius. His contradictions didn't bother us and we accepted that his philosophy was one of aphorisms and not a system. He made up the Overman, he intoxicated through his language and we were elevated far above the ordinary by his bold thoughts.[83]

Obviously, Jung's encounter with *Zarathustra* wasn't the solitary experience he remembered but was, in fact, one he shared with young people all over Europe. Nietzsche's writings fired the imagination of an entire generation that heeded his call to create a new cultural order in Europe. He invited them to cast aside old habits of mind and dare to live their life with the same commitment with which he had lived his. During the 1890s Nietzsche became the locus of a cult fascinated with his madness, one that seemed to be a virtual reenactment of Dionysos' dismemberment. As Jung later recalled the question remained "Did Nietzsche embody the very theme of degeneration that he had railed against?" One way to answer this in the affirmative was to join one of the creative movements that sprang up at the time. In general, they shared a *Lebensphilosophie* ("life-philosophy") that emphasized the role of the irrational and myth in revitalizing a culture debased by the philistine tastes of the middle classes. It appealed to a new intelligentsia that rejected the dominant materialist philosophy in order to explore various paths that ranged from theosophy to ritual magic and mysticism.

The World of Art

In art and literature naturalism was rejected in favor of symbolism. Artists such as Gustav Moreau disdained the superficiality of an "impressionist" treatment of everyday life and found inspiration in the inner world of

Imagination for his fantastical treatment of classical mythology. Another early Symbolist was the Basel-born painter Arnold Böcklin (1827–1901) who had studied with Burckhardt and was inspired by him to study in Italy after a stay with the Düsseldorf School of landscapists. Nymphs and satyrs populated his bucolic paintings that at their best created a vivid mood evoking the living presence of the antique world. His later paintings such as *The Isle of the Dead* were more self-consciously symbol-laden, a fact that only seemed to enhance their popularity. Jung referred to it during his *Zarathustra Seminar.*

> So the analogy which Nietzsche uses here is partially a speech metaphor or a poetic image, and partially due to primitive reasons. The land of the dead is often an island—the island of the blessed, or the island of immortality, or the isle of the graves where the dead are buried or the ghosts are supposed to live... So Nietzsche's picture of the silent isle in the ocean is quite true to type, and he has to sail over the sea to reach that place where the dead live. You have probably seen the picture called "The Island of the Dead" by our famous Swiss painter Böcklin; it is practically everywhere in the form of picture post cards and such horrors.[84]

The leading Swiss Symbolist writer was another Baseler Carl Spitteler (1845–1924). Like Böcklin, he was deeply influenced by the artistic sensibilities of his teacher Burckhardt. His reputation rests on two ambitious epics *Prometheus and Epimetheus* (1881) and *Olympian Spring* (1900–1906). His work figured in the development of psychoanalytic theory when the title of his novel *Imago* was appropriated as an early conceptual term. Jung analyzed *Prometheus and Epimetheus* in his lengthy chapter on poetry in *Psychological Types.*[85] Both Böcklin and Spitteler turned to classical mythology for inspiration but chose lesser-known figures that they treated in a contemporary idiom. Jung would adopt this approach in his own line of work.

This distinct milieu of fin de siècle Basel also left its mark on an aspiring young German writer named Hermann Hesse who came to the city in 1899 to work in a bookstore. He was introduced to local society by Rudolf Wackernagel, an old friend of his father and an archivist and city historian. In particular, Hesse availed himself of the city's artistic resources, meeting Heinrich Wölfflin who had assumed Burckhardt's chair at the university. He was able to frequent the art museum where he could visit the Böcklin room and admire the work of his favorite painter. "You know," he wrote, "how I have always adored Böcklin even before I had seen any of his work in the original."[86] Hesse shared this neo-romantic sensibility with Jung and would later undergo Jungian analysis, first with Josef Lang in 1916 and later with Jung himself.

Hesse signed his first book contract in 1899 with the Eugen Diederichs Verlag in Jena. As Germany's leading publisher of neo-romantic literature, this prestigious firm brought out ten works by Spitteler. To Diederichs

> every book was an individual work of art with its own unique cultural mission. He believed that not only the written content of the book, but also its external form and design must carry a cultural message to the public. Each book had to be given a soul of its own, a unique form which distinguished it from all others. This outward aesthetic form, however, must be in complete "organic" harmony with the book's content.[87]

With this goal in mind, Diederichs championed *Jugendstil*, the German version of Art Nouveau. Characterized by curving lines and rich ornamentation, this art movement spread through Europe during the 1890s and dominated the decorative arts until World War I. Jung's library contained many books from Diederichs Verlag, which also published numerous works on Romantic nature-philosophy and history of religions (e.g., gnosticism and the translations of Richard Wilhelm).

Since artistic training was an integral part of the classical education that Jung received, we need to consider his formative art experiences. In *MDR* Jung remembered back to his boyhood the parlor at Klein-Hüningen.

> Here all the furniture was good, and old paintings hung on the walls. I particularly remember an Italian painting of David and Goliath. It was a mirror copy from the workshop of Guido Reni; the original hangs in the Louvre. How it came into our family I do not know. There was another old painting in that room which now hangs in my son's house: a landscape of Basel dating from the early nineteenth century. Often I would steal into that dark, sequestered room and sit for hours in front of the pictures, gazing at all this beauty. It was the only beautiful thing I knew.[88]

One can feel here Burckhardt's pervasive influence both in his enthusiasm for Renaissance Italy and for his loyalty to the German landscape tradition. Like all students Jung was required to master the fundamentals of art and his sketchbooks attest to his talent for architectural details. He struggled through his art classes at the Gymnasium.

> I had some facility in drawing, although I did not realize that it depended essentially on the way I was feeling. I could draw only what stirred my imagination. But I was forced to copy prints of Greek gods with sightless eyes, and when that wouldn't go properly the teacher obviously thought I needed something more naturalistic and set before me the picture of a goat's head. This assignment I failed completely, and that was the end of my drawing classes.[89]

After receiving his medical degree in 1900 Jung painted a number of watercolor landscapes that showed he had not permanently abandoned his artistic efforts. They are clearly inspired by the German landscape tradition, examples of which hung in the Jung household and in many others around Basel. One can discern the specific influence of Carl Gustav Carus (1789–1869), the pioneering gynecologist who is best remembered for *Psyche* (1846), which Jung recognized as a precursor to his own theory of the unconscious. Carus also became an accomplished landscapist, receiving advice from Goethe and the friendship of Germany's greatest Romantic painter Caspar David Friedrich. He developed his theory of art in *Nine Letters on Landscape Painting* (1815–1824) where he said that landscape painting "must express a state of mind... this can be so only where the natural landscape is apprehended and depicted from an aspect that coincides exactly with the inner mood in question."[90]

The work of Jung's that most successfully achieved this objective was a twilight scene he painted in 1901–1902 before going to Paris to study with Pierre Janet. A stream descends past a pine and three poplars to a marshy valley at twilight.[91] It delicately captures that time of day's "magical" mood, a time that had attracted the interest of the first generation of Romantics with their fascination for liminal experiences. "Twilight" symbolized the borderline between day-consciousness and night-consciousness. At the same time he was demonstrating a talent for an *artistic* expression of this phenomenon, Jung was, as a psychiatrist, studying it as a *psychological* state.

At the time Jung visited the city, Paris had clearly established its ascendancy as Europe's cultural capital, hosting the 1900 International Exposition that showcased the technological and cultural achievements of the Age of Progress.[92] Jung went to the Louvre and attended the theater during his two sojourns there. To his cousin Helene who was learning dress-making there he wrote "If you don't have time, then I suggest next Sunday evening at 7 1/2 in front of the Sarah Bernhardt theatre. They will probably be doing *Theriogne de Mericourt* which is very beautiful. *Resurrection* at the Odeon is rather grander and *Le Joug* at Mme. Rejane's funnier. I'll go along with whatever you wish."[93]

Just walking the streets exposed Jung to the Art Nouveau movement then at its height. He would pass the recently built Metro stations with their sinuous railings designed by Hector Guimard and scan the advertising pillars covered with a riot of colorful posters competing for the attention of passers-by. Among the aspiring artists who achieved their first fame and fortune as graphic designers was the Czech Alphonse Mucha who was chosen by Sarah Bernhardt to be her official poster designer. She fell in love with his work and also had him design jewelry, costumes, and sets for her. Mucha's signature style of an ornately dressed central figure surrounded by intricate decorative detail was widely imitated to help sell everything from bicycles to insurance.

Bernhardt and Mucha shared an interest in Byzantine art. One of Bernhardt's greatest roles was in *Theodora* written by Victorien Sardou. He also wrote for her the 1894 *Gismonda*, a tragic romance whose sets included a Byzantine church.[94] To prepare herself for *Theodora*, Bernhardt visited Ravenna to study the mosaics in the Church of San Vitale.[95] Ravenna was the capital of the Byzantine Empire's Italian province for several centuries and is filled with churches built with imperial largesse. The most impressive of these was San Vitale where the mosaic retinues of the emperor Justinian and his wife Theodora face each other across the apse. Bernhardt did sketches of Theodora that became the designs for the costume and jewelry that cost a small fortune. In 1903 Gustav Klimt also visited there and was so impressed by the mosaics he saw that he developed the deliberately Byzantine style for which he became most famous.

As we shall see, the pictures Jung painted in his Red Book after his 1914 visit to Ravenna show the degree to which his artistic sensibility was transformed by his encounter with Byzantine art. In reminiscences cut from *MDR*, Jung commented on his difference of opinion with Burckhardt regarding Byzantine art.

> It is amazing what Burckhardt saw and didn't see in Italy. He couldn't relate to Ravenna but that was due to changes in taste. Goethe hadn't seen Giotto: this is a psychological prejudice that accompanies secularization. That is the style that also underlies genius, it is the loudspeaker of its era. Burckhardt was narrow in his judgement: this incapacity to grasp Ravenna.[96]

Jung here identified an important shift in taste occurring in late-nineteenth-century German-speaking Europe. The younger generation was enthusiastically taking up the call of *Zarathustra* to live life to the fullest. This meant treating life as a total work of art and led to the call for a *Gesamtkunstwerk* ("total artwork") that included not just the fine arts but such applied arts as interior decoration and furniture. Young people looking for meaningful alternatives to the bankrupt Christianity of their parents' generation were not content with Burckhardt's stoic adherence to the cultural canon of "old Europe." Nietzsche invited each of them to treat personality as a work of art. The very passions that upset Burckhardt about Nietzsche were the key to his appeal and dominant influence on the emerging *Lebensphilosophie* ("life philosophy") movement.

Jung's cultural views were shaped by Burckhardt and Nietzsche, Basel's two icons. He adopted their pessimistic assessment of developments in modern society but went on to articulate his own unique program for renewal.

This therapy was based on the deep personal experiences of art and religion that he had while growing up in Basel as well as his attraction to the neo-romantic movement popular during his university years. Younger intelligentsia considered themselves to be the avant-garde of a new European cultural community although many of them moved in a more distinctly conservative direction as they grew older.[97] Jung can be counted among them; he would take an active role in the "cultural wars" that were waged in German-speaking Europe from the Wilhelmine Period through the Cold War. Ideologically, he would eventually find a group of Swiss and German conservatives to be his most congenial network.

Chapter 2

Freud and the War Years

Interested in a wide range of scientific and philosophical topics Jung faced a major decision in choosing which medical specialty to pursue. Friends were surprised when he declined an invitation from Friedrich von Müller, professor of internal medicine at the University of Basel, to accompany him to Munich.[1] Jung opted instead for the then poorly regarded field of psychiatry and joined the staff of the Burghölzli, Zurich's cantonal hospital and a center for the innovative treatment of mental illness. It was headed by Eugen Bleuler among whose many contributions to the field was the introduction of terms such as "autism" and "schizophrenia" (a condition then called "dementia praecox"). He had established a therapeutic community that attracted the most talented young psychiatrists of Europe and America. Jung soon became Bleuler's chief assistant and quickly distinguished himself in the field with a book on dementia praecox and a series of publications on his word association experiments. This pioneering work in experimental psychopathology established the existence of emotionally charged elements in the unconscious that he referred to as "complexes."

Of the many new treatments under consideration none was more passionately discussed than that of the controversial Viennese neurologist Sigmund Freud. A.A. Brill later recalled the atmosphere at the hospital this way "It was inspiring to be in a group of active and enthusiastic workers who were all toiling to master the Freudian principles and to apply them to the study of patients. Psychoanalysis seemed to pervade everything there."[2] Jung eagerly joined this discussion and began to adopt many of Freud therapeutic techniques.

Jung went to Vienna in early 1907 to meet Freud. The visit led to a deep personal bond that fostered their mutual interest in probing the deeper recesses of the human mind. To better understand their patients the men studied dreams and used hypnosis, techniques that were academically suspect. Jung became psychoanalysis' chief spokesman and helped organize a series of psychoanalytic congresses at Salzburg (1908), Nuremberg (1910), Weimar (1911), and Munich (1913) that laid the organizational framework for the movement.

Jung's doubts about psychoanalytic theory began in 1909 and eventually resulted in *Wandlungen und Symbole der Libido* (*Transformations and Symbols of Libido*) (1912), his sprawling intellectual declaration of independence from Freud. He went a step further when he began to record his own fantasies pictorially as well as in writing in what became known as the *Red Book*.[3] This avant-garde side was tempered during the war by Jung's adoption of decidedly right-wing views on contemporary politics and culture.

Joining Forces with Freud

Although born a generation apart, both men came to their careers in the natural sciences with an education steeped in the neo-humanistic curriculum of the German-speaking world. Their boyhood imaginations were whetted by Greek mythology and further stimulated by a series of archeological discoveries that dramatically altered the study of history. The most famous of these was Schliemann's discovery of Troy. His excavations proved the historical basis of the Homeric epic and provided a potent metaphor for the work the two men were undertaking in the new field of depth psychology (another term coined by Bleuler).[4]

Their collaboration was articulated in a vocabulary derived from the terminologies of the new human sciences established during the nineteenth century. Besides archeology, they kept abreast of developments in anthropology and history of religions; Jung had chosen psychiatry as a profession as a way to reconcile his interest in both the natural and the human sciences. He was an ambitious young man who threw himself into their project with enthusiasm. As time went on, he began to feel frustrated with Freud's inability to accept his modifications of the libido theory that stemmed from his work with schizophrenics. He had become convinced that an exclusively sexual interpretation could not explain the material that emerged in deeply regressed psychotic states. Jung's reservations had a more personal side since he felt that Freud was also incapable of understanding the personal dreams he shared on their trip to America in 1909. "As Freud could only partially handle my dreams, the amount of symbolical material in them increased as it always does until it is understood. If one remains with a narrow point of view about the dream material, there comes a feeling of dissociation and one feels blind and deaf. When this happens to an isolated man he petrifies."[5]

For Jung, this impasse was best exemplified by the dream of a house that he had on the return voyage. He first related it in his 1925 seminar on Analytical Psychology and later in a different version in *MDR*.[6] In it he descended down through the different floors of the building until he reached its deepest underground level where he saw prehistoric bones and pottery. To Jung, Freud's exclusive interest in the skulls as expressions of unconscious death wishes missed the dream's true significance. "My dream constituted a kind of structural diagram of the human psyche; it postulated something of an altogether *impersonal* nature underlying that psyche."[7]

Jung's dedication to psychiatry and psychoanalysis had left him little time for his former intellectual pursuits. His move to Zurich had been a conscious effort to make his own way in the world. What has escaped notice is the degree to which the Basel milieu, which he thought he was leaving behind, resurfaced during this time of psychic dislocation.

The house dream, for example, evoked the medieval architecture of his old hometown, and the city figured more explicitly in another dream of the period that began with his encounter with an Austrian customs official.[8] After a hiatus, the dream shifted to an Italian cityscape like Bergamo that reminded Jung of the Kohlenberg ("Cabbage Hill"), a neighborhood in Basel whose streets are partly flights of steps leading down to the Birsigtal, a small river valley. It is the place where the river flows into a system of sewers dating back to the Middle Ages that flow under the city and finally empty into the Rhine. Jung would have walked those steps often on his way between town and Bottminger Mill where he had moved with his mother and sister after the death of his father in 1896.

In the dream, one stairway leads down to the Bärfüsserplatz. The square got its name from the Franciscan ("barfüsser" = "barefoot") Church there that had been converted into the city's historical museum. The pride of its collection was the remnants of the cathedral treasury that had been divided between Basel-Stadt and Basel-Land in 1833 with one-third being awarded to the City and the rest to the Land.[9] Jung was making his way through a summer noonday crowd and saw a knight in armor covered with a Crusader tunic. For Jung, he symbolized a particular state of consciousness.

> I had grown up in the intensely historical atmosphere of Basel at the end of the nineteenth century, and had acquired, thanks to a reading of the old philosophers, some knowledge of the history of psychology. When I thought about dreams and the content of the unconscious, I never did so without making historical comparisons.[10]

He initially associated the knight with his teenage interest in stories about the Holy Grail, which he later amplified to include alchemy. Having reached an impasse with the Freudian model of the psyche he now began to study the world's symbolic systems, a project that was to last the rest of his life.

After his return from America in 1909, Jung immersed himself in mythology and archeology. He wrote Freud that he "was reading the 4 volumes of old Creuzer, where there is a mass of material. All my delight in archeology (buried for years) has sprung into life again."[11] This led to his writing *Wandlungen und Symbole der Libido*, which was published in 1912. Its opening trope clearly announced this shift.

> The impression made by this simple reference may be likened to that wholly peculiar feeling which arises in us if, for example, in the noise and tumult of a modern street we should come across an ancient relic—the Corinthian capital of a walled-in column, or a fragment of inscription. Just a moment ago we were given over to the noisy ephemeral life of the present, when something very far away and strange appears to us, which turns our attention to things of another order; a glimpse away from the incoherent multiplicity of the present to a higher coherence in history.[12]

This theme of the reanimated past is pervasive during this period. In one dream Jung had encountered a knight walking the streets of Basel, in another he walked past a series of mummies from different historical periods who come back to life under his gaze. "Dreams like this, and my actual experience of the unconscious, taught me that such contents were not dead, outmoded forms, but belong to our living being."[13] He now conceived of the psyche at its deepest level as a network of dynamic, imaginative patterns. Although he later designated them as "archetypes" Jung initially called these patterns "primordial images," a term he took from Jacob Burckhardt.[14]

As he came to realize how saturated his own personal fantasies were with historical material, Jung began to articulate his new model of the psyche in a vocabulary influenced by the humanistic *Bildung* he had received in Basel. For Jung, his new approach was to be counted among the "human" and not the "natural" sciences (the *Geisteswissenschaften* rather than the *Naturwissenschaften*). He eventually drew on literature, anthropology, and the history of religions to elucidate his theories since he now realized that the psyche was culturally scripted and not biologically determined. To study the psyche he employed the comparative methodology he had learned at the university. As discussed in the previous chapter, this method had its origins in the morphological studies of Goethe, and had been continued by a group of German scientists opposed to mechanistic natural science. "It has become quite clear to me," Jung wrote in a letter to Freud (December 25, 1909) "that we shall not solve the ultimate secrets of neurosis and psychosis without mythology and the history of civilization, for *embryology* goes hand in hand with *comparative anatomy*, and without the latter the former is but a freak of nature whose depths remain uncomprehended."[15]

For Jung the key breakthrough was extending the comparative method from the anatomy of the body to that of the psyche. He wrote to Freud that

> Antiquity now appears to me in a new and significant light. What we now find in the individual psyche—in compressed, stunted, or one-sidedly differentiated form—may be seen spread out in all its fullness in times past. Happy the man who can read these signs! The trouble is that our philology has been as hopelessly inept as our psychology. Each has failed the other.[16]

"Although the philologists moan about it, Greek syncretism, by creating a hopeless mishmash of theogony and theology, can nevertheless do us a service: it permits reductions and the recognition of similarities, as in dream analysis."[17] As his letters indicate, Jung hoped that Freud would find his exuberance contagious: "We are on the threshold of something really sensational, which I scarcely know how to describe except with the Gnostic concept of *sophia* [in Gk], an Alexandrian term particularly suited to the reincarnation of ancient wisdom in the shape of ΨA [shorthand for psychoanalysis]."[18]

After their break, Jung was explicit about the hermeneutical intent of his new psychology. "Our position is more like that of an archeologist deciphering an unknown script."[19] A dream was not to be interpreted as a disguise but as an unknown text whose pictorial language needed to be translated. Like a philologist who had to know different languages, the analyst would need to be familiar with the history of symbols to better understand the message contained in a particular dream. Jung was here articulating his new understanding of the unconscious and how it differed from that of Freud. He felt that Freud's focus on neurotic symptomatology was ignoring other important discoveries in medical psychology.

The Vienna/Zurich Divide

Partisan politics have colored memories and influenced accounts about the split that was to divide the psychoanalytic movement. Even at the time participants were sensitive to the quasi-racial vocabulary being used to describe and usually deride psychoanalysis; it is now widely agreed that Freud's adoption of Jung as his "crown prince" was part of his plan to insure that psychoanalysis would avoid being labeled "a Jewish national affair" and become an accepted part of mainstream science. Freud's success in "conquering" the Burghölzli created hurt feelings among his original Viennese followers who resented their loss of status.

This element was mostly confined to private communications prior to the 1914 publication of Freud's *On the History of the Psycho-Analytic Movement* in which he wrote that Jung had "seemed to give up certain racial prejudices which he had previously permitted himself."[20] There is a certain irony here since the use of racial vocabulary was more evident in Freud's camp than in Jung's, receiving its most extended treatment in the letters Freud wrote to Karl Abraham. In an early one he referred to the fact that their "racial kinship" made it easier for Abraham to accept his theories than Jung who "as a Christian and a pastor's son" came to psychoanalysis after "great inner resistances."[21] Freud would later employ "Aryan" as an alternate designation for their erstwhile Zurich colleagues. He was careful to use a more culturally neutral term when he was writing to Jung, "I find the racial mixture of our group most interesting; he [Ernest Jones] is a Celt and consequently not quite accessible to us, the Teuton and Mediterranean man."[22] The two men did discuss the ethnic dichotomy more directly during their 1909 trip to America. In a letter to his wife Jung wrote "Freud and I spent several hours walking in Central Park and talked at length about the sociological problems of psychoanalysis... We spoke a good deal about Jews and Aryans, and one of my dreams offered a clear image of the difference."[23] No other account of what exactly the two the men spoke about has surfaced. Despite the fact that all the early psychoanalysts spoke about Aryans and Jews Saul Rosenzweig singled out Jung for having employed a "racist dichotomy," an assertion typical of the anti-Jung bias to be found in psychoanalytic historiography.[24]

Ernest Jones noted "how extraordinarily suspicious Jews could be of the faintest sign of anti-Semitism and of how many remarks or actions could be interpreted in that sense."[25] Even after opinions on both sides had hardened, all Fritz Wittels, an early follower of Freud, could manage to say on the subject was that "It is probable that the Swiss were not entirely free of race prejudice."[26] What *was* the nature of Swiss anti-Semitism? To Abraham it seemed to be that of "a certain type of German."[27] In a letter to him Freud described it as "suppressed" but acknowledged that it was directed at Abraham in more overt fashion.[28] Of course, it would also be possible to interpret Abraham's observation in the context of the long Swiss tradition of guarded tolerance of foreigners. Remember that in 1897 Basel had hosted the first Zionist Congress. At the Burghölzi, a nondiscriminatory policy meant that the brightest young psychiatrists from around the world were invited to join the staff. Any slights that Abraham may have experienced there would have stemmed more from the social insularity of the Swiss than from any overt anti-Semitic hostility on their part.

Jung did criticize the Viennese, but not for their being Jewish; they were, in his view, "a degenerate and Bohemian crowd" (an opinion that was shared by Freud).[29] Adopting Burckhardt's view of the assimilated Jew as the "agent of modernity" he was alienated more from their atheism rather than their ethnicity.[30] He countered their Enlightenment critique of religion with a response that reflected the period's avant-garde enthusiasm for the history of religions.

He decried the "poverty of symbols" that had begun with the Judeo-Christian hostility to idolatry.[31] He felt that this condition was being perpetuated by the psychoanalytic reliance on the positivistic premises then prevailing in science. He chose to articulate his differences in a terminology derived in part from the alternative movements popular at the time, which emphasized the creative potential of unconscious forces; Jung's designation of his new methodology as "psychosynthesis" was dismissed by Freud as "Aryan religiousness."[32]

In discussing his own religious identity Freud made blatantly contradictory remarks. To Abraham he wrote that "we Jews lack the mystical element" but to Jung he wrote about the "specifically Jewish nature of my mysticism."[33] Jung could not abide Freud's doubts about the religious function of the psyche; if he was to be Freud's Joshua, he wanted to follow the Moses of the Burning Bush, not the Moses of the Ten Commandments. Jung was looking for personally transforming experiences and had no appetite for what he felt was the growing psychoanalytic preference for an overly narrow, legalistic approach to their new field.

Jung's ideas about "race" were derived for the most part from the contemporary preoccupation with the *Volksseele* ("national soul") found in such works as Le Bon's *Psychological Laws of the Evolution of Peoples* (1894). This is first made clear in *Wandlungen und Symbole der Libido* where he made public his new conception of the libido.

> *There must be typical myths which are really the instrument of a folk-psychological complex treatment.* Jacob Burckhardt seems to have suspected this when he once said that every Greek of the classical era carried in himself a fragment of Oedipus, just as every German carries a fragment of Faust...A way is here opened to the understanding of secret springs of impulse beneath the psychologic development of races. (Emphasis in the original)

Finally, in language that has a Bergsonian flavor,

> The unconscious is the generally diffused, which not only binds the individuals among themselves to the race, but also unites them backwards to the peoples of the past and their psychology...Man as an individual is a suspicious phenomenon, the right of whose existence from a natural biological standpoint could be seriously contested, because from this point of view, the individual is only a race atom, and has a significance only as a mass constituent.[34]

Jung's psycho-spiritual understanding of race was in contrast to the "scientific" view propagated by the racial-hygiene movement.[35] Fin-de-siècle concerns about "degeneration" had led to numerous studies that purported to demonstrate the "feminine inferiority" of Jews, the best-known of which was Otto Weininger's *Sex and Character*. Sander Gilman incorrectly links Jung to this development when he says that in 1913 "Freud's discourse about blond Aryans such as Jung and their eternal opposition to the 'dark' Jews framed his conflict with Jung. He simply reverses the rhetoric of race applied to him by Jung."[36] This argument is flawed since Gilman does not, and cannot, quote anything that Jung wrote about Freud at that time but has to rely on things he wrote twenty years later.

When Jung publicly discussed Freud it was from a psychological rather than a racial point of view. He developed his new theory of psychological types in part to understand the differences that were quickly splitting the so-recently-formed psychoanalytic movement. "On the one side we have [Freud's] theory which is essentially reductive, pluralistic, causal, and sensualistic...On the other side we have the diametrically opposed theory of Adler, which is thoroughly intellectualistic, monistic, and finalistic."[37] These remarks were made in the paper he gave in Munich in 1913 at the last psychoanalytic conference he was to attend. Over time, however, this neutral formulation would give way to an increasingly negative evaluation. First Jung came to lump them together as "leveling" psychologies that focused only on human shortcomings and then, in a 1934 letter, he sarcastically labeled them a "Jewish gospel" of an "essentially corrosive" [*wesentlich zersetzenden*] nature.[38]

Privately things were otherwise. In 1912 Freud wrote to Ludwig Binswanger

> how far away [the Zürichers] must have gone from the understanding of the Ucs. which is our pride, if like our most simple-minded opponents, they want to drag in racial differences. The only serious thing about it is this: Semites and Aryans or anti-Semites, whom I wanted to bring together in the service of psi analysis, once again separate like oil and water.[39]

Preoccupied with his mythological studies Jung was uncommunicative so Freud took up the ethnic issue with Alphonse Maeder. In October Maeder wrote that

> the Semitic/Aryan mentalities (*Weltanschauungen*) are different and I believe they complement one another...our entire enterprise is marked by the Semitic spirit in a manner adverse to adjustment and that we should be conscious of it...I think it is time to realize this state of affairs, since it is our duty as analysts to go to battle as unprejudiced as possible...[40]

The exchange continued the next year when Freud wrote Sandor Ferenczi that on the matter of Semitism "there are certainly great differences from the Aryan spirit. We can be convinced of that every day. Hence there will be differences of world views and art here and there. But there should not be a particular Aryan or Jewish science."[41]

The final rupture came when Jung wrote to Freud on October 27, 1913, "Maeder tells me you doubt my *bona fides*."[42] Several weeks later Jung vented his anger in a letter to the Swedish analyst Poul Bjerre. "Until now I was no anti-Semite, [but] now I'll become one, I believe."[43] When Freud shortly afterward published *On the History of the Psycho-Analytic Movement* and publicly labeled him one Jung's anger only increased. It turned into a grudge that was to color his feelings about Freud and his former psychoanalytic colleagues for the rest of his life.

Analytical Psychology Takes Shape

Jung's immersion in mythological material led him to seek clinical proof of the psyche's phylogenetic level. He had his assistants Sabina Spielrein and Johann Honegger collect data from their patients who were diagnosed with introversion psychoses. Each provided examples of "archaic thought traces," the most famous being the delusions of one of Honegger's cases who is remembered as "the Solar Phallus Man."[44] His delusion about a tube hanging down from the sun that was a source of the wind bore an uncanny resemblance to a passage in some ancient magical texts thought by Albrecht Dieterich to be the remnant of an authentic Mithraic liturgy. For Jung this provided crucial validation of his theory that mytho-poetic components existed and were still operative in the modern psyche.

Richard Noll goes to great length to catalog Jung's misconduct regarding Honegger's papers and legacy but as is so often case, Noll's criticism misses an important point. His account ignores Jung's 1912 visit to Saint Elizabeth's Hospital in Washington, D.C. where he collected material from a group of Negro patients. For Jung it provided important cross-cultural verification for his hypothesis.

> In order to settle [the question of the inheritance of mythological images] I went to the United States and studied the dreams of pure-blooded Negroes, and I was able to satisfy myself that these images had nothing to do with so-called blood or racial inheritance, nor are they personally acquired by the individual. They belong to mankind in general, and therefore they are of a *collective* nature.[45]

In July 1914 the Zurich branch withdrew from the International Psychoanalytical Association and shortly afterward renamed itself the Association for Analytical Psychology. Jung was now leader of the Zurich School and went about getting the group a publisher. He secured a contract with Deuticke who was already the official publisher of the psychoanalytic movement. In his foreword to *Psychologische Abhandlungen* (*Psychological Papers*) Jung announced that they went beyond the boundaries of psychopathology to investigate issues of a general psychological nature. There were contributions by four of his associates, the two most important being Josef Lang and Hans Schmid. Along with Maeder, Lang was on the staff of the sanitorium of Dr. Bircher-Benner the creator of Bircher-muesli, the health cereal. He soon joined the staff of the Sonnenmatt Sanatorium in Lucerne where he analyzed Hermann Hesse. Schmid and Jung became very close; they became godparents to each other's children and traveling companions to northern Italy in 1914 where they visited Ravenna. Their lengthy correspondence about psychological types helped Jung clarify his thinking on the subject.[46]

The Zurich School was finding an audience in America where Jung's 1912 Fordham lectures *The Theory of Psychoanalysis* were published by The Nervous and Mental Disease Monograph Series, which also brought out works by Riklin on fairytales and Maeder on dreams. A milestone was reached in 1916 with the publication of *Psychology of the Unconscious*, Beatrice Hinkle's translation of *Wandlungen und Symbole der Libido*. Hinkle had received her medical degree in San Francisco and moved to New York in 1905 where she joined the staff of Charles L. Dana, America's leading neurologist, at the Cornell University Medical College. She developed an interest in psychoanalysis and went to Europe for two years (1910–1912) where she met Freud and Jung and attended the Weimar Conference. Upon her return, she took up residence in Gramercy Park and rejoined the Cornell faculty. She seems to have been responsible for Jung being invited to lecture to the Liberal Club during his stay in New York in the spring of 1913. The Club was located near Gramercy Park and had begun as the Public Forum started by Percy Stickney Grant (1860–1927), pastor of the Episcopal Church of the Ascension in 1907, as a place to discuss contemporary social issues. Among the speakers at its weekly meetings were Booker T. Washington and Margaret Sanger, the founder of the birth-control movement. Hinkle introduced Jung to many of her friends in Greenwich Village. He met Kahlil Gibran who drew his portrait and dined with members of the Heterodoxy Club, the first feminist group in the United States.

Jung spoke to the Liberal Club in March and later that year it went through a crisis. Henrietta Rodman, a public school teacher with radical views, led a revolt against the more moderate "social settlement" group from the church. She was instrumental in the club's relocation to MacDougal Street in Greenwich Village where it became the center for political radicalism, sexual liberation, and artistic innovation. Nearby were the editorial offices of *The Masses*, most of whose writers favored Jung, and The Provincetown Playhouse. Eugene O'Neil later said of *Psychology of the Unconscious* that "If I have been influenced unconsciously, it must have been by this more than any other."[47]

One of O'Neil's collaborators, the set-designer Robert Edmond Jones, later went to Zurich to analyze with Jung. His experience there reinvigorated his enthusiasm for the theater and led to his writing *The Dramatic Imagination,* which explained his application of Jung's approach to creative fantasy.[48] Jung's book was widely reviewed and influenced Jack London in his last collection of stories based on traditional Polynesian mythology, *On a Makaloa Mat.*

Jung was also attracting adherents in England who preferred his approach to the unconscious to that of Freud. The most important was Dr. Constance Long who edited the *Collected Papers on Analytical Psychology,* which appeared in 1916. Jung's introduction served as a manifesto for his group.

> The Zurich School has in view the end-result of analysis, and it regards the fundamental thoughts and impulses of the unconscious as symbols, indicative of a definite line of future development... Out of the symbolic application of infantile trends there evolves an attitude which may be termed philosophic or religious, and these terms characterize sufficiently well the lines of the individual's future development.[49]

Although Long died shortly after joining Beatrice Hinkle in New York in 1923 she inspired another Englishwoman Esther Harding to follow. Harding joined Eleanor Bertine and Kristine Mann who together with Beatrice Hinkle and school psychologist Frances Wickes formed the nucleus of the Jungian movement in the United States.

The Zurich School's New Interests and Venues

During these years Jung was also busy promoting psychoanalysis at home as well as abroad. At the end of 1911 his article "New Paths in Psychology" appeared in *Raschers Jahrbuch für Schweizer Art und Kunst.* He also got involved in a controversy about psychoanalysis that raged in the popular press and is discussed in great detail by Henri Ellenberger.[50] Besides writing letters to the *Neue Zürcher Zeitung,* he continued his defense of psychoanalysis in an article written for the literary journal *Wissen und Leben.* These venues are of importance because they indicate some of the new outlets Jung was finding for his work (he would continue to contribute to *Wissen und Leben* after it was renamed the *Neue Schweizer Rundschau*). The appearance in 1917 of his *Die Psychologie der unbewussten Prozesse* in a monograph series by Rascher was the beginning of his formal affiliation with that publisher, whose ambition was to offer in his pages a "true mirror of Helvetic intellectual life." The Jahrbuch sponsored the work of "the younger generation," featuring literature by such figures as Hesse and Spitteler.

In 1916 Rascher published Maeder's study of the painter Ferdinand Hodler. Hodler (1853–1918) was born in the Berner-Oberland, lived in Geneva, and gained international fame with his Symbolist masterpiece *Night* (1890). Hirsch has demonstrated that after the 1896 Swiss National Exposition he returned to the Swiss subjects of his early years but with a different palette and vision. Awarded the commission to paint the Hall of Weapons in the new National Museum, he did so against intense opposition from the director. It depicted *The Retreat from Marignano*, the crushing defeat in 1515 that ended Swiss active participation in European affairs. He celebrated the robust monumentality of his native land in his landscapes, historical subjects such as William Tell, and portraits of such notables as Spitteler and General Wille, the commander-in-chief of the Swiss army during World War I. "Hodler painted his conservative, Swiss subjects in an inventive, internationally influenced style."[51]

Maeder's study of Hodler was the first to apply the Zurich School's new insights about the creative role of unconscious fantasy to works of art.[52] In Maeder's opinion, Hodler's work expressed the national ideals of intensity and clarity and was a true synthesis of the Germanic and French components of the Swiss *Volkspsyche*. This synthesis mirrored a federal system that balanced the interests of the individual cantons with those of the nation as a whole. At the deepest levels of fantasy creative individuals such as Hodler encounter transpersonal psychic factors and become mouthpieces for trends found in the group to which they belong.

The Zurich School found an ally in the Institute of Psychology and Psychotherapy (Geneva), which espoused an eclectic approach. Several of its members such as Edouard Claparede were at the university and had known Jung for years. Another member who would later maintain a long and sympathetic relationship to Jung was Charles Baudouin (1893–1963). In his *Le Symbole chez Verhaeren* (translated into English as *Psychoanalysis and Aesthetics* [1924]) he analyzed the works of the Belgian poet Emil Verhaeren using Maeder's study of Hodler and Jung's recently published *Psychological Types*. He also translated Spitteler into French.

When considered in the context of the neo-romantic artistic and philosophical trends of the period in which it was created Jung's psychology no longer stands out as the deviation it is often labeled. An analysis of the book catalogue for Jung's personal library makes clear his decided preference for Romantic literature. Among the Romantic classics he owned were Chateaubriand's *Atala* and Nerval's *Aurelia*; other examples of historical romance are Haggard's *She*, Benoit's *L'Atlantide*, and Erskine's *The Private Life of Helen of Troy*.[53] Jung also relied on two poetic epics in writing the books that established his professional reputation. He used Longfellow's *Hiawatha* for amplifying Frank Miller's fantasies in *Wandlungen und Symbole der Libido* and, as previously mentioned, Spitteler's *Prometheus and Epimetheus* for *Psychological Types*.

Jung took a keen interest in the neo-romantic revival of fairy tales and the occult. In an introduction to a book written by his follower Oscar A.H. Schmitz, Jung wrote "The content clothed itself in fairytale form not with the secret pretence of being an allegory, but because in this guise it could find the simplest and most direct access to the reader's heart."[54] Writers such as Ernst Barlach, Alfred Kubin, and Gustav Meyrink wrote tales that explored the dark, irrational dimension of reality and Jung cited all of them in *Psychological Types* to support his theories. Gustav Meyrink (1868–1932) lived in Prague and then moved to Munich where he worked for the popular cultural journal *Simplicissimus*. His preoccupation with occult themes is evident in his two most famous novels *The Golem* (1915) and *Das grüne Gesicht* [*The Green Face*] (1916). Jung discussed them in his 1925 and 1928 English language seminars and owned a number of Meyrink's other books. Kurt Wolff remembered being unable to convince Jung that *Das grüne Gesicht* was a bad novel as he thought highly of it. Jung did not judge a work by its literary merit but by the degree to which he felt it was inspired from the collective unconscious.[55]

Maurice Maeterlinck (1862–1949) was another literary figure with an interest in fairy tales and the occult that Jung began to read at this time. After arriving in Paris from Belgium he became a leader in avant-garde Symbolist theater and took up residence in a former abbey where he staged many productions. *Pelleas and Melisande* (1892) inspired a composition by Debussy and its success was to be surpassed by that of *The Blue Bird* (1909); two years later he won the Nobel Prize for literature. Something written about him also applies to the avant-garde conservatism that characterized Jung. "It is curious that a man who is so modernistic in mind...should place all his dramas in the historical legendary past."[56] His mystical leanings were expressed in various works on the spirit world, plants, and animals. In *Wandlungen und Symbole der Libido* Jung used Maeterlinck's concept of the "inconscient supérieur" to support his new views about the prospective function of fantasy.[57]

Jung's preference for "visionary" over "psychological" art led him to oppose modernist experimentation with its rejection of ornamentation and historical references, its preference for fragmentation, and its celebration of meaninglessness. He would make clear his antipathy for what he considered the nihilistic trends in such modernist icons as Dada, Picasso, and *Ulysses*. "I loathe the new style, the new Art, the new Music, Literature, Politics, and above all the new Man."[58] Dada was the first of this "new Art" to become the object of his scorn. In a 1918 article he wrote "This lost bit of nature seeks revenge and returns in faked, distorted form, for instance as a tango epidemic, as Futurism, Dadaism, and all the other crazes and crudities in which our age abounds."[59] Ernest Jones remembered "I recollect asking [Jung] once whether he thought the vogue of Dadaism, just then beginning in Zurich, had a psychotic basis. He replied: 'It is too idiotic for any decent insanity.' "[60] Jones identified 1908 as the year that this was said but this must be an error since Dada only began in 1916 at the Cabaret Voltaire; this would suggest that Jones stayed in touch with Jung after his withdrawal from the psychoanalytic movement in 1914. This may have occurred in 1919 when he was in Zurich to marry a second time.

Like Spitteler Maeterlinck was published by the Eugen Diederichs Verlag, one of Germany's leading publishers of neo-romantic literature. Gary Stark documents Diederichs' indebtedness to Burckhardt and Nietzsche and his role in promoting authors opposed to materialism in philosophy and naturalism in literature.[61] Jung relied on other EDV authors such as Arthur Drews and Albert Kalthoff to support his views on the mythological basis of Christianity. C.A. Bernoulli, a follower of Ludwig Klages, was a fellow Baseler who wrote a book about Nietzsche and Overbeck for Diederichs, which Jung quoted in *Wandlungen und Symbole der Libido*.[62] Jung later became personally acquainted with several of the firm's authors; in the early 1920s he analyzed Hesse and befriended Richard Wilhelm and Leopold Ziegler. All these new literary interests with their emphasis on symbols and the inner life indicate the direction Jung was moving in after his break with Freud.

The Schwabing Connection

As Jung severed his relationship with Vienna (he didn't visit the city again until 1928) his connection to Munich, the site of the 1913 psychoanalytic conference, was growing stronger. This had less to do with the local analysts Leonhard Seif (who had analyzed his wife Emma) and Hans von Hattingberg than with his interest in Schwabing, the city's bohemian quarter and Germany's countercultural mecca.[63] His entree into that milieu was Otto Gross whom Jung had known both as a patient and as a psychoanalytic collaborator. In 1908 Gross was hospitalized at the Burghölzli for opium addiction where he was treated by Jung. The analysis became a mutual one, Jung writing to Freud that "Whenever I got stuck, he analyzed me. In this way my own psychic health has benefited too."[64] He soon reported that "Gross, unguarded for a moment, escaped over the garden wall and will without doubt turn up again in Munich, to go towards the evening of his fate."[65] That fate included being diagnosed a schizophrenic by Jung, waging a protracted legal battle with his father over his guardianship, and dying of pneumonia in 1920.

Years later Jung recalled that Gross "mainly hung out with artists, writers, political dreamers, and degenerates of any description, and in the swamps of Ascona he celebrated miserable and cruel orgies."[66] Ascona, a village in the Italian part of Switzerland, had become the destination of choice for the European avant-garde, attracting pioneers in modern dance as well as various faddists and political anarchists. Jung's comment belies the influence Gross had on his newly liberated attitude toward women. In 1909 Sabina Spielrein wrote in her diary that Jung "arrives beaming with pleasure, and tells me with strong emotion about Gross, about the great insight he has just received (i.e., about polygamy)..."[67] About their relationship Jung wrote to Freud that "I have learnt an unspeakable amount of marital wisdom, for until now I had a totally inadequate idea of my polygamous components despite all self-analysis."[68] In the background was Bachofen's theory of mother-right, which Gross, like many in Schwabing, championed as an alternative to the rigidly patriarchal ethos of Wilhelmine Germany.

The blurring of boundaries between the personal and professional reflected the popularity of a *Lebensphilosophie* that promoted authenticity as a way to overcome the artificial constraints imposed by modern society. The exasperation felt by the older generation toward this development was succinctly expressed by Friedrich von Müller, Jung's old medical school mentor, who declared that the work of Gustav Richard Heyer, one of his students who later became Jung's German lieutenant, was "not science, but Schwabing!"[69] As a young man Heyer had belonged to the Stefan George Circle and was attracted to the biocentric philosophy of Ludwig Klages who had been conducting psychodiagnostic seminars at Munich University since 1903. This influence can be seen in his 1932 book *The Organism of the Soul*, published by Lehmanns Verlag, Germany's leading publisher of medical books and major promoter of racial hygiene and other völkisch causes.[70]

The cultural ferment of prewar Schwabing nurtured another of Jung's Weimar-era followers Oscar A. H. Schmitz (1873–1931). After reading a poem by Hugo von Hoffmansthal in the *Blätter für die Kunst* he was inspired to become a lyric poet; he joined the Stefan George Circle, wrote symbolist poems, and befriended Meyrink. Schmitz was close to Fanny von Reventlow the "queen of Schwabing" who counted Klages among her many lovers.[71] He also became close to the artist Alfred Kubin who married his sister Hedwig in 1904. Although challenged by Berlin after the turn of the century, Munich continued to be the artistic capital of Germany. In December 1900, at a time Kandinsky was studying at the atelier of Franz von Stuck, Jung made a post-graduation trip to the city where he visited the collection of Old Masters at the Alte Pinakothek.[72]

Like Böcklin, von Stuck painted many nymphs and satyrs but went on to make his fame and fortune with such crowd-pleasers as "Sin." It depicted a voluptuous nude woman with a serpent curled around her that he painted in multiple versions. He dominated the Munich art scene that Jung was beginning to frequent. In 1909 Jung wrote to Freud that he had spent a week there and "gorged" himself on art.[73] Two years later, after a rendezvous there with Freud he stayed on and, following Freud's example, bought an oil painting and three drawings.[74] In 1912 he wrote that he had spent the New Year's holiday "traveling breathlessly around Germany visiting various art galleries and improving my education."[75]

Jung, Magian Culture, and the Feminine

Jung used this art to amplify the theory of creative fantasy he was in the process of developing in *Wandlungen und Symbole der Libido*. He also showed an interest in Byzantine art that was attracting widespread interest at the time. The mystical spirituality of Eastern Christianity appealed to sensitive souls troubled by the Industrial Revolution and the utilitarian standardization that followed in its wake. In a tone reminiscent of Blake, Kandinsky warned that "The nightmare of materialism, which has turned the life of the universe into an evil, useless game is not yet past; it holds the awakening soul still in its grip."[76]

Jung shared Kandinsky's rejection of Western materialism and like many found alternatives in what Oswald Spengler was to identify as "Magian Culture." This cultural zone had stretched from Spain through the Middle East and Central Asia to China. Its prime symbol, the world-cavern, was reflected in the domed and ornately decorated architecture of Byzantium and Islam. Immersing himself in the newly available Gnostic writings, Jung sought to enter this world through his imagination. His personal "Journey to the East" left traces in the *Red Book* and in *Seven Sermons to the Dead*, which he wrote under the pseudonym "Basilides in Alexandria." He continued these explorations in the 1920s with his trip to North Africa and his friendships with Count Keyserling and Richard Wilhelm. All this reflects his place in the neglected field of German Orientalism, which has been extensively studied by Suzanne Marchand.[77]

The decisive experience in Jung's artistic development was his visit to Ravenna where he was most impressed by the tomb of Galla Placidia. Because of its protected location, the city became the capital of the Western Roman Empire and its successor states. Its rulers spent lavishly on church construction, making the town home to many masterpieces of early Byzantine art. Although the tomb is overshadowed by the nearby church of San Vitale, its modest exterior gives way to an interior space completely covered with brilliant mosaics dominated by shades of blue and gold. Jung remembered it as "significant and unusually fascinating."[78]

The mosaics had already inspired such individuals as Bernhardt and Klimt who visited in 1903. A friend of Klimt's recalled that "the gleaming gold mosaics in the churches of Ravenna made a tremendous, decisive impression on him. From then on its magnificence, its frozen splendour was a feature of his sensitive art."[79] Jung also adopted a Byzantine aesthetic when he began his active imaginations. The best-known is his picture of Philemon who had first appeared to him in a dream.[80] This figure "brought with him an Egyptian-Hellenistic atmosphere with a Gnostic coloration."[81] A winged Philemon is dressed in a long robe decorated with floral patterns and reverentially holds a light in his cupped hands. He hovers over a domed building beside which are a knotted serpent and grove of date palms; above him are three rondels against a deep blue background. Although Jung would have been familiar with axial composition, ornate costuming, and the halo effect from Mucha posters, two specific elements were inspired by his Ravenna visit. The first is his treatment of the rondels, which closely resemble those found on the ceiling of the tomb of Galla Placidia; the other is the domed building, which is modeled on the Tomb of Theodoric.

Many of the other figures Jung painted are costumed in a distinctly Orientalist style, dressed in pantaloons and slippers in settings done in boldly colored geometric tilework and arabesques.[82] They closely resemble the costumes and sets of the productions of the Ballet Russe such as *Scheherezade* (1910). Jung owned *Tibetan Paintings* (1925) by George Roerich whose father Nicholas was the set designer for *The Rites of Spring*. Robert Edmond Jones who did the sets for the company's 1916 American production of *Till Eulenspiegel* was struck by the theatricality of the *Red Book* pictures when Jung showed them to him during the course of his analysis.[83] Other pictures in the *Red Book* show his appreciation of tribal, Celtic, and Meso-American art that he got from his travels and collection of ethnographic books.

In contemplating his decision to begin practicing active imagination Jung was filled with a great deal of fear and resistance. In spite of the uncertainty, he felt compelled to proceed and was astonished when a woman's voice, that of a psychopathic patient, said that it was "art." He could not bring himself to agree with her and decided that it was neither "science" nor "art" but "nature." His reason for doing so was the conviction that these personifications of his fantasy were not personal creations but autonomous expressions of the collective unconscious.[84] This crisis of creativity influenced Jung's formulation of his archetype of the anima who is a condensation of various female cultural figures and individuals in a man's life.

Although he considered Galla Placidia an embodiment of his anima, the most important such figure to emerge in his active imaginations was Salome who appeared in the company of Elijah who then evolved into Philemon.[85] John Kerr points out that Jung avoided mentioning the two people in his life who most embodied the Logos/Eros dyad at that moment in his life, namely Freud and Sabina Spielrein.[86] This is consistent with Jung's emphasis on the impersonal nature of his fantasies; he associated the pair with Simon Magnus/Helen,

Klingsor/Kundry, and Lao-Tzu and the dancing girl. His treatment of the biblical references was superficial and missed the obvious association of Elijah as the prophet who opposed King Ahab after his wife Jezebel, a Phoenician princess, had introduced the cult of Baal.

Salome was a minor biblical figure but *the* major fin-de-siècle icon of the femme fatale, triggering a veritable Salomania. She paraded through the art, literature, and music of the period from Moreau to Klimt and von Stuck, from Huysmans to Wilde, Beardsley, and Strauss (even young Picasso drew her).[87] Her popularity crossed the Atlantic and sparked a "Salome craze" that was in full swing when Jung and Freud visited the United States. Salome's "Dance of the Seven Veils" helped inaugurate modern dance as well becoming the "hoochie-coochie" that gave rise to strip-tease. Robert Henri's paintings of Salome represented her as one of these early strippers. In a letter to his wife from New York Jung described a night out on the town. "Next we went to a real Apache music hall, a rather gloomy place. A singer performed, and the audience showed its appreciation by throwing money on the floor at his feet."[88] The term "Apache halls" originated in Paris to describe the variety show venues that appealed to a rowdy, lower class audience. "Apache" was meant to convey the air of wild freedom associated with Geronimo and his warriors of the Southwest who had gained notoriety eluding the U.S. Army's efforts to capture them. One wonders if a "Dance of the Seven Veils" was also on the bill the night Jung was there.

Jung was strongly attracted to Jewish women, which he described as his "amiable complex" in a letter to Freud about Spielrein who "was systematically planning my seduction."[89] He then recalled his infatuation with another Jewess while at an Adriatic resort that he visited with his wife after their stay in Vienna in 1907 and had had a similar experience while studying in Paris.[90] This attraction might stem from a generational crush on Sarah Bernhardt who was half-Jewish and stirred the imaginations of several generations of European males. Jung was to transfer his affections from Spielrein to another young patient Antonia Wolff who was half-Jewish and with whom he had an intimate relationship that lasted until her death in 1953. She became an analyst and influenced the course of analytical psychology through her analytical work and writing.

One of her most important theoretical pieces was "Structural Forms of the Feminine Psyche" in which she discussed a quartet of female psychosocial types: mother, Hetaira, medial woman, and Amazon.[91] The sources of these terms illuminate the intellectual milieu that Jung and Wolff inhabited. The first two were adapted from the Penelope/Calypso dyad discussed by Hans Blüher in his book *The Role of Eros in Male Society* (Diederichs, 1910). He created a scandal by asserting that homosexuality was a normal, even preferable, behavior. "Blueher was a fervid anti-feminist, and wrote with great frankness about his own erotic inclinations as a youth (later, he did marry and have a family)."[92] The dyad encapsulated the alternate roles of wife or companion that were available to women of the time, an issue that was a passionately debated one among feminists and members of the German youth movement.

> Ascona was characterized by the presence of remarkable women who were *not* political feminists. The first of these is Fanny von Reventlow. In 1899 she published in Oscar Panizza's *Züricher Diskussionen* her essay entitled "Viragines oder Hetaerae?" in which she repudiated the women's movement, and defined herself as a hetaera (the modern equivalent is perhaps "free woman").[93]

"Hetaira" and "Amazon" were terms coined by Jacob Bachofen in his writings on prehistoric matriarchy. The fourth term "medial woman" was adopted from the book *Femmes inspiratrices et poetes annonciateurs* by the occultist Edouard Schuré (Perrin & Co., 1907). Jung navigated his midlife crisis by reaching an understanding with his wife Emma about Toni's place in their marriage. The marital compromise was one in which monogamy was amended to accommodate Jung's Orientalist harem fantasy. Taking his cue from Gross he soon became more comfortable with polygamy and told Freud that "The prerequisite of a good marriage was the license to be unfaithful."[94]

We can now return to Salome and consider other aspects that reveal just how overdetermined her appearance in his fantasies was. The original Elijah preached an uncompromising allegiance to Yaweh and opposed the polytheistic idolatry introduced by Jezebel. In this reading the "Druidic sacred place" that figured in the active imagination discussed in his 1925 Analytical Psychology Seminar would be based on his familiarity with the contest between Elijah and the prophets of Baal at the altar on Mount Carmel told in the First Book of Kings, Chapter 18. Yaweh looked with favor on his sacrifice and the crowd turned on the four hundred and fifty losers who were marched to a stream where Elijah cut their throats. Jung remembered that he "had the feeling of diving into an atmosphere that was cruel and full of blood."[95]

Another cultural polarity relevant here was that between the feminine (Galla Placidia-Salome-Jezebel) decadence of Oriental Culture and the masculine vigor of the German barbarians. During the nineteenth century, standard historical accounts taught that the Roman Republic had conquered the East only to succumb to the allure of its exotic religious cults. The resulting hybrid empire was so racially and spiritually enfeebled that it proved no match for the invading German tribes that crossed its borders and established their kingdoms. Jung identified himself with the barbarian Ataulf who took Galla Placidia as his wife. He was the brother of Alaric, conqueror of Rome, and became king of the Visigoths after Alaric's death. In connecting Galla Placidia to his concept of the anima Jung wrote that

> She provides the individual with those elements that he ought to know about his prehistory. To the individual, the anima is all life that has been in the past and is still alive in him. In comparison to her I have always felt myself to be a barbarian who really has no history—like a creature just sprung out of nothingness, with neither a past nor a future.[96]

Jung's imaginative identification with his barbarian heritage would also have a source in the völkisch books he acquired in his youth. One was *Die Ahnen* ("The Ancestors," 1872) by Gustav Freytag who wrote *Debit and Credit* (1855) a best-seller in Germany that was responsible for promoting the stereotype of the post-emancipation Jew as a rootless being bent only on getting rich. Another was *Tuisko-Land, the Aryan Race and Divine Homeland* (1891) by Ernst Krause, which discussed his theory of the original homeland of the Aryans and their later migrations. One final book in this genre from Jung's library is *Hypatia, or New Foes with an Old Face*, a novel by the Victorian writer Charles Kingsley. A proponent of muscular Anglo-Saxon virtues, he wrote in the book's preface about "those Gothic nations of which the Norwegians and Germans are the purest remaining stock" and how "the races of Egypt and Syria were effeminate, overcivilized, exhausted by centuries during which no infusion of fresh blood had come to renew the stock" (London: Macmillan and Co., 1902). As we shall see, Jung would identify himself as a "German" in an important 1918 paper and contrast them with Jews.

Ellenberger described this as a period of Jung's "creative illness" in which he responded to all the stress in his life with an incredible burst of creativity. His effort to express his latent potentials can be seen as his personal *Gesamtkunstwerk* (total work of art). From his active imaginations to his relationships Jung sought to realize his generation's Nietzschean credo that said life itself was the highest form of art. Life was to be a performance rather than the object of merely intellectual analysis.

> So our way has to be one where the creative character is present, where there is a process of growth which has the quality of revelation. Analysis should release an experience that grips us or falls upon us as from above, an experience that has substance and body, such as those things occurred to the ancients.[97]

For Jung Elijah was a shaman whose mana conveyed a sense of godlikeness symbolized by his presence at the Transfiguration. In a passage deleted from *Wandlungen und Symbole der Libido* Jung wrote "We who are reborn again from the mother are all heroes together with Christ and enjoy immortal food."[98] This situation created the danger of identification with the collective unconscious but Jung resisted Salome's advances when she tried to worship him as Jesus Christ. (Noll based much of his thesis on a misreading of this experience. He claimed that Jung's experience caused him to imagine he had undergone divinization and thus entitled to preach a new religion.)

Jung's break with Freud precipitated a crisis that led to his distinct approach to therapy that fostered a creative encounter with fantasy material. The following quotes show the extent to which Jung described his new psychology in the spirit of *Lebensphilosophie.*

> I think we must give it time to infiltrate into people from many centres, to revivify among intellectuals a feeling for symbol and myth, ever so gently to transform Christ back into the soothsaying god of the vine... what infinite rapture and wantonness lie dormant in our religion, waiting to be led back to their true destination![99]

"The right interpretation for a symbol (analytical or constructive, cf. *The Content of the Psychoses,* 2nd Edition) is the one that brings out the greatest value for life (a pragmatic view)."[100]

What Jung found most objectionable about Freud was his effort to disenchant an already disenchanted world with views that were "a sinful violation of the sacred."[101] To Hans Schmid he wrote,

> The symbol wants to guard against Freudian interpretations, which are indeed such pseudo-truths that they never lack for effect. With our patients "analytical" understanding has a wholesomely destructive effect, like a corrosive or thermocautery, but it is banefully destructive on sound tissue. It is a technique we have learnt from the devil, always destructive... In the later stages of analysis we must help people towards those hidden and unlockable symbols, where the germ lies hidden like the tender seed in the hard shell.[102]

The War

The assassination of Archduke Franz Ferdinand of Austria on June 28, 1914, led to a summer of tension and ultimatums. On August 2 Germany invaded neutral Belgium as part of its plan to deliver a knock-out blow against France. Jung was in Scotland for a conference and it took a week of traveling through Holland and Germany for him to get home.

> I came right through the armies going west, and I had a feeling that it was what one would call in German a *Hochzeitsstimmung*, a feast of love all over the country. Everything was decorated with flowers, it was an outburst of love, they all loved each other and everything was beautiful. Yes, the war was important, a

> big affair, but the main thing was the brotherly love all over the country, everybody was everybody else's brother, one could have everything anyone possessed, it did not matter.[103]

This "spirit of 1914" and the national community it briefly seemed to embody was to be remembered by critics of the Weimar government who would invoke it as the model for the new society they hoped to inaugurate.

Switzerland's army of 250,000 was mobilized on August 3 and stayed on active duty until the end of hostilities in 1918. Pledged to the defense of Swiss neutrality it had been professionalized along Prussian lines by Ulrich Wille who was made commander-in-chief. Since it was a citizen's army just about every Swiss man had to serve and this caused financial hardship for many. Jung served with the medical corps and eventually became the commandant of an internment camp for Allied officers at Château d'Oex.

The violation of Belgian neutrality divided Swiss public opinion along linguistic lines with the French- and German-speaking regions sympathetic to opposing belligerents. With passions running high Carl Spitteler was invited to deliver a speech to the Zurich branch of the New Helvetic Society, an organization founded in February to discuss topics of national interest. In "Our Swiss Standpoint" the distinguished man of letters reminded his audience exactly what was at stake for the country. He reminded them that no matter how strong their feelings for Germany, loyalty to their fellow countrymen was their patriotic duty. Its timeliness was underscored by another publication from Rascher Verlag—*We Swiss, Our Neutrality and the War.* It was an anthology with contributions from such regulars as C.A. Bernoulli, Robert Faesi, and Adolf Keller. Emil Ermatinger, a professor of German literature at the ETH (Federal Technical University) and the University of Zurich, was a new contributor and would later edit a literary anthology that included Jung's "Psychology and Poetry."[104]

A group of Swiss intellectuals including Ferdinand Hodler issued their "Geneva Protest" against the two most barbaric acts committed in the war's opening phase: the German destruction of the library at Louvain University in Belgium and its bombardment of Rheims Cathedral. Hodler was immediately condemned in Germany and his mural at the University of Jena covered up while Spitteler was denounced for the criticisms of Germany he had expressed in his neutrality speech.[105]

These pleas were directed at a Swiss-German population whose sympathy for Germany was so strong that it included passing military intelligence to the German army. Many belonged to such organizations as "The German-Swiss Society" and "The League of Overseas Germans." They had studied at German universities while Germans were prominent in Swiss business affairs and intellectual life. Ferdinand Sauerbruch was a German who became a professor of surgery at the University of Zurich and director of the surgical clinic of the cantonal hospital. When war broke out he went back to Germany but returned to Zurich where he became famous for developing a prosthetic hand for those crippled in combat. The Swiss Eugen Bircher, a colleague of Jung's in the army's medical corps, was the chief surgeon at the Aargau cantonal hospital. During the war he served as a Red Cross doctor with the German army on the Bulgarian front. When the war was over he founded the "Swiss Fatherland Association" and became active in right-wing politics. A colleague of Bircher's, Emil Sonderegger came into the public eye in 1912 when he coordinated the army maneuvers in east Switzerland that were attended by the Kaiser. He is best remembered for commanding the troops that suppressed the general strike in Zurich in November 1918. The specter of an imminent Bolshevik revolution had created panic among the middle class of Central Europe and the Swiss government took no chances. A man of extreme right-wing views, Sonderegger became active in so-called Front organizations in 1933.

What does all this have to do with Jung? Quite a lot since it provides the background for strong views that Jung held but which dropped from sight in later years. The image most people have of Jung during World War I involves his premonitions of a pending bloodbath and possibly his assignment at Château d'Oex. A careful reading of his writings helps fill the lacunae. The most important discovery is the fact that Jung condoned the German invasion of Belgium.

> When they broke into Belgium they said yes, we have violated the Treaty; it *is* mean. That is what Bethmann-Hollweg always said; "We *have* broken our word," he confessed. And then we said how cynical he was and that the Germans were only pagans anyway. But they simply admit what the others think and do.[106]

Jung accepted the German rationalization for their violation of Belgian neutrality and found it preferable to the hypocrisy of the other belligerents, especially England, which was routinely criticized by German intellectuals for its unlimited capacity for "cant." Jung's sense of *Realpolitik* can also be seen in his reaction to the wartime destruction of cultural monuments. Complaining to Spielrein in 1917 he wrote "With what contempt people have treated the libido work and intellectually torn it to shreds! They have bombarded it intellectually, but it is nevertheless quite clear that a gothic cathedral and a library of old manuscripts are nothing in the face of the thoroughly decisive power of a 28-cm. shell."[107]

Jung's main preoccupation was how the war affected his inner life. In spring 1914 he had a dream in which Europe was in the grip of an Arctic cold wave that had a positive outcome when it occurred the third time. "There stood a leaf-bearing tree, but without fruit (my tree of life, I thought), whose leaves had been transformed by the effects of the frost into sweet grapes full of healing juices. I plucked the grapes and gave them to a large, waiting crowd."[108] This sense that he had a therapeutic mission to accomplish was also evident in another of his wartime dreams.

> In the beginning of the war I was always dreaming of having interviews with Kaiser Wilhelm, and I always tried to convince him that he should retire with his royalties, but he would never listen. We knew each other quite well; when I appeared, he used to wave at me, and I said, "Yes, I am here again and I have to tell you that you should retire!"... it was useless, you see. I did not succeed at all. It stopped in the end of 1916.[109]

While Jung was at Château d'Oex he drew a mandala every day in a notebook.[110] He had done his first one in 1916 after completing *Seven Sermons*. Mandala means "sacred circle" in Sanskrit and is symbolized by a circle, square, or quaternion, which was developed in Tibetan Buddhism into an elaborate cosmology. For Jung it represented psychic totality, the self beyond the ego that is the goal of the individuation process. The mandala became the dominant motif in Jung's late active imaginations. He later painted a luminous gold sun with a red cross and border in the *Red Book*.[111] It blazes with stylized flames that radiate into Signac-like dabs of color. Hovering beneath the mandala is a yogi sitting on a carpet, his eyes closed in meditation holding a vase above his head. The lower half of the picture is rare in Jung's work on account of its folk-art portrayal of a Swiss landscape. A fortification manned by soldiers looks over a rifle-range and a road leading past a canal and railroad yard to a walled city with ships sailing in the background. The city recalls the fantasy he had as a boy and that was discussed at the beginning of chapter one. The picture's upper plane represents the sacred world of transcendent spiritual totality with the yogi acting as the mediator between it and the mundane world below. Jung has divided this everyday world into contrasting views of Switzerland. On the left is the rural world of his youth with its farm animals, fields, and sailboats; on the right across the road is the modern industrial world of factories, railroads, and steamships. A strong wind blows from left to right to keep the pastoral side from being polluted by the smoke of modernity.

The War's Aftermath

In 1918 as the guns on the Western Front fell silent Jung's article "The Role of the Unconscious" was published in *Schweizerland: Monatshefte für Schweizer Art und Arbeit* ("Switzerland: Monthly for Swiss Style and Work"). Founded by a group of Rascher authors it appeared from 1914 to 1921 when it was reorganized as the *Schweizer Monatshefte* and became one of the country's leading conservative journals. He first presents a survey of the scientific study of the unconscious from Janet and Freud to his own findings about the supra-personal or collective dimension of the human mind. It was characterized by mythological fantasies (soon to be called "archetypes") that were created by the activity of the brain itself. Jung made it clear that he was referring to *innate possibilities* along the lines of Kantian categories rather than actual inherited images. He used anthropological examples to illustrate his points to remind his readers that the only advantage they had over their primitive fellow man was their greater linguistic facility since otherwise they both shared common psychological experiences.

At this point his argument takes a sharp turn and he begins his first published discussion of the psychological differences between Germans and Jews. This is important because it is the basis of the better-known and more controversial remarks he made in 1933–1934. A review of the original manuscript in the Jung Archive at the ETH (Hs 1055: 27) reveals that an opening diatribe about Jews and Aryans was not included in the published version. More importantly, the English translation found in the Collected Works (Volume 10), its first, is seriously flawed by a number of significant deletions and interpolations. The first is the deletion of "Germans" in the following sentence. Freud's "specifically Jewish doctrines are thoroughly unsatisfying to the Germanic mentality: we ["wir *Germanen* (my italics)"] still have a genuine barbarian in us who is not to be trifled with..." This deletion necessitated tampering with a paragraph that preceded it. The English reads "As civilized human beings, we in Western Europe have a history reaching back perhaps 2,500 years." In German: "Wir haben als Kulturmenschen ein Alter von etwa Funfzehnhundert [1500] Jahren." To his original German-speaking readers, the "we" was a clear reference to themselves as Germans and the "1500" would refer back to the time when the Germans were converted to Christianity. In the English version the "we" is supplemented with "in Western Europe," which necessitates recalibrating the time span from 1,500 to 2,500 years ago to include the ancient Greeks.

Jung's entire argument was structured in terms of the cultural stereotypes about Aryans/Germans and Jews that were current at the time. Scholars have emphasized the major shift that occurred after the popularization of the theory of evolution when Jews were no longer seen as a religious group but rather as a distinct race with identifiable physical and mental traits.[112] Jung accepted this and some of the basic characterizations derived from it. One was the contrast of the "rootedness" of the Aryan people and the "rootlessness" of Jews ("where has he his own earth underfoot?" [par. 18]).

Jung continued his discussion in terms familiar to his German-speaking audience. "The Jew already had the culture of the ancient world and on top of that has taken on the culture of the *nations amongst whom he dwells*" ["*Wirtsvolk,*" my italics] (par. 18, my italics). The English translation glosses over a nuance that is highly significant. "Wirtsvolk" is better translated as "host people" and had become commonplace in discussions about the relationship of the Jews to the larger, national communities around them. Given this linguistic premise, there were two possible definitions associated with Jews. Most Europeans considered them "guests" while others, influenced by the racial hygiene movement, labeled them "parasites." In any case they were "aliens" separated from their Aryan neighbors.

One of Jung's intentions in writing this article was to articulate his compensatory theory of the unconscious as a complement to Freud's repression theory. For Jung, compensation was one of the basic features of psychic functioning. Analogous to the body's homeostatic system, it balances the one-sidedness of conscious awareness with such unconscious material as dreams and symptoms. Since Jews had insufficient contact with the earth and the world of instincts, he found it understandable that Freud and Adler would reduce everything to its material beginnings. This became Jung's basic critique of Freud and would gain popularity with those uncomfortable with Freudian psychology.

> The fact is, our unconscious is not to be got at with over-ingenious and grotesque interpretations. The psychotherapist with a Jewish background [more accurately, "The Jewish-oriented psychotherapist"] awakens in the Germanic psyche not those wistful and whimsical residues from the time of David, but the barbarian of yesterday, a being for whom matters become *serious* in the most unpleasant way.[113]

A careful reading of the article also reveals an important linguistic strategy that he repeated elsewhere in his writings. In paragraph 19 he writes first about "the specific Jewish need to reduce..." and then, in referring to Freud and Adler, to "these specifically Jewish doctrines." This was Jung first use of the designation "specific," which he would later use as a standard qualification of Jewish thought. He used it in a manner that exuded certainty and aimed to close rather than initiate a discussion.

Jung is making clear his identity as a German in the *Kulturkampf* that had began to unfold during the war over the "ideas of 1789." The war had led to growing polarity between the Left and Right that increasingly included the issue of "the Jewish problem." His criticism of Freud focused not so much on his biological membership in a "Jewish race" but on his atheistic, materialistic premises. Freud the assimilated Jew was an "agent of modernity" dedicated to disenchanting rather than re-enchanting the world. Shortly after the suppression of the 1919 Spartacist uprising in Berlin and the assassination of its leaders Karl Liebknecht and Rosa Luxemburg Jung wrote to Spielrein "What has

Liebknecht to do with you? Like Freud and Lenin, he disseminates rationalistic darkness which will yet extinguish the little lamps of understanding."[114]

These opinions found validation in the anti-Semitic views of the White Russian émigré Emil Medtner who had become his patient during the war. He had been active in the Symbolist movement and was obsessed with the destructive influence of Jews on modern culture. Besides his analytical relationship, he became a personal friend of Jung's. A founding member of the Psychology Club Zurich, he was the man responsible for the translation of *Wandlungen und Symbole der Libido* into Russian, which was taken to Russia by a group of Mensheviks returning for the revolution. "East versus West and Aryan versus non-Aryan were prominent topics in their dialogue."[115] This move to the right can also be seen in the fact that Jung began to cite Leon Daudet's "eminently readable *L' Heredo* [*Heredity: an essay on the interior drama*]."[116] Daudet's theory about the spontaneous appearance of "ancestral units" in the personality shows the influence of Le Bon. Daudet (1867–1942) was from a distinguished French literary family. As a medical student he had been a friend of Charcot's son Jean and met Freud when they were both dinner guests there in 1886. He became a royalist and fanatical anti-Semite who helped found the proto-fascist organization *L'Action Francaise.*

Just how public Jung's antipathy to the Left was by this time can be seen in the epilogue of his recently completed *Psychological Types.*

> In our age, which has seen the fruits of the French Revolution—"Liberté, Egalité, Fraternité"—growing into a broad social movement whose aim is not merely to raise or lower political rights to the same general level, but, more hopefully, to abolish unhappiness altogether by means of external regulations and egalitarian reforms—in such an age it is indeed a thankless task to speak of the complete inequality of the elements composing a nation.[117]

In the years after the war Jung was to grow close to a group of German intellectuals who had fought on the war's cultural front. The most important was Oscar Schmitz who was remembered as "this type of German-Jewish mixture who was very nationalistic and an officer in the German army in World War I. He had much national political interest as a writer also."[118] In 1915 he published *The Real Germany* in which he proclaimed that a "New German Man" had been born in 1914. He rebutted Allied propaganda that claimed the war was a contest between "Civilization" and "Barbarism" seeing it as a struggle between Western "Civilization" and German "Kultur" and declared that the German idealistic concern for "spirit" and "soul" was superior to Western materialism. This theme was found in many other intellectual contributions to the war effort such as Rudolph Eucken's *The Moral Power of War,* Thomas Mann's *Reflections of an Unpolitical Man*, Max Scheler's *The Genius of War and the German War*, Werner Sombart's *Merchants and Heroes*, and Leopold Ziegler's *The German Man*. "The Role of the Unconscious" can be seen as Jung's contribution to this school of literature. As a member of the conservative wing of the avant-garde Jung sympathized with their opinions and would affiliate himself with critics of the Weimar Republic who felt that it was an alien imposition on German national life.

Chapter 3

Jung's Post-Freudian Network

With the war over Jung first became active in the Anglo-American world where he began to attract interest after the appearance of *Psychology of the Unconscious* and *Collected Papers on Analytical Psychology* in 1916. His trip to the American Southwest (1924–1925) was a result of a long-standing relationship with the McCormick family of Chicago, while his trip to British East Africa (1925) was planned by H.G. Baynes. Through his New York circle Jung was invited to speak at several international education conferences in Switzerland, England, and Germany. After the German situation stabilized around 1925 he began to more actively promote his work there by lecturing at the School of Wisdom and to the Kulturbund. He became part of a network of neoconservative intellectuals there who were interested in breathing new life into a cultural agenda whose rationale had been called into question by the war. The conservative subtext of Jung's writings of this period has been lost but is *the* crucial background for understanding the better known and more controversial views that he was to express in later years.

The Anglo-American Connection

At first the British physician Maurice Nicoll seemed the natural candidate to lead a Jungian group in London; during the war he wrote *Dream Psychology*, a book that showed a sympathetic understanding of Jung's departures from Freudian theory. It was at a summer house in Buckinghamshire rented by Nicoll that Jung had his famous spook experience while on his first postwar trip to Britain in 1919. Unfortunately for Jung Nicoll soon became interested in the Russian spiritual teacher Gurdjieff and moved to his headquarters at

Fontainebleau.[1] H. Godwin Baynes became the de facto leader and together with Esther Harding, Kristine Mann, and Eleanor Bertine formed the first generation of Jungian analysts. Since there was no formal training institute until 1948, the primary method of training, besides personal analysis with Jung, was participation in a series of seminars that Jung first began to hold in the United Kingdom in the 1920s. The first two were held in Cornwall: at Sennen Cove in 1920 and at Polzeath in 1923. Attendance grew from a dozen to over one hundred at the third held at Swanage in southern England in 1925.

The Polzeath seminar was entitled "Human Relationships in Relation to the Process of Individuation." In it Jung began the process of articulating the core concepts of his new theory of the psyche. This included his theory of the collective unconscious and its structural components the archetypes. He also made frequent references to psychological types, his most famous contribution to practical psychology. Interest would have been lively since his book on the subject had just appeared in an English translation by Baynes. His discussion ranged over his now familiar medley of dreams, symbols, the transference, and what was then called "the psychology of primitives." His most sustained analysis was of what he called "the four exclusions of Christianity." These involved the repression of nature, animals, primitives, and creative fantasy. He discussed the reasons for these repressions, the consequences, and how these repressions manifest themselves (e.g., in such cults as those of the body and of pets).[2]

One of the most important features of Jung's exposition, and a constant in his subsequent writings, was contrasting his approach to psychology with that of Freud. This had begun in his article "The Psychology of Unconscious Processes" (1917) where he characterized the theories of Freud and Adler as reductive with their one-sided emphases on Eros and the Will to Power, respectively.[3] In these seminars Jung made a further, significant distinction by postulating a qualitative difference in the dreams of Jews and people of Germanic stock due to differences in the psychic development of each group.

To illustrate this Jung gave two examples from his analytic practice. The first dream was that of an elderly professor from a Catholic university. In it, he was up in the Alps and came across a balustrade made out of Greek marble upon which a naked woman with the feet of a chamois was dancing. This evidence of classical civilization led Jung to conclude that this was a "Jewish dream." After getting angry, the man did confess to Jewish parentage. The second involved a pedagogue in an old German family who fell into a psychogenic state that had started as a dream. He was walking on a hillside with rabbit holes all around. Pottery and implements were scattered about and he discovered more artifacts after some digging. Jung declared that this was a Germanic dream since the man had dug and found the primitive. These conclusions were based on Jung's understanding of the historical development of the two groups. The Germanic peoples were forcibly converted by the Romans to Christianity, which was grafted onto the stump of their old religion. When the unconscious of a Germanic person is probed, evidence of their primitive heritage is immediately apparent. A Jew, on the other hand, has incorporated the antique civilization of the Mediterranean into his psychic makeup; he does not possess the intensity of the primitive, because it is already dissolved into the antique world. Jung said that Freud was right to uncover sexuality from the point of view of Jewish psychology. To the Jew it is necessary to discover his sexuality, while people of Germanic stock are already aware of it and want to know what to do with it.

Jung repeated the Jewish dream in his 1925 Swanage seminar on dream analysis and expanded on his opinion of Freud and his method. He observed that Freud did not take the general conventions seriously because he was a Jew and so had the law in his veins. As the inheritors of antique civilization, the instincts of Jews were worn out. If one uncovered such things as the fire in the Jews, they welcomed it and were not afraid. Such a man as Freud was not threatened in the least. His ideas would, however, uncover something in Jung and his audience (being of Germanic stock), which should not be uncovered. They would be morally smashed while the Jew welcomed any trace of instinct since he was already petrified like a nearly extinct volcano.

These comments resonated with many who heard Jung since they are corroborated by two individuals with whom he was on close terms, the German philosopher Count Hermann Keyserling and the British psychologist William McDougall, about whom I will have more to say. Keyserling wrote that "C.G. Jung has shown, by a comparison of the dreams of Jews with those of Christians, that at the same level of the subconscious where the Germanic type is still a lake-dweller, the Jew is an Alexandrian."[4] McDougall wrote

> each race and each people that has lived for many generations under or by a particular type of civilization has specialized its "collective unconscious," differentiated and developed the "archetypes" into forms peculiar to itself...He [Jung] claims that sometimes a single rich dream has enabled him to discover the fact, say, of Jewish or Mediterranean blood in a patient who shows none of the outward physical marks of such descent...He points out that the famous theory of Freud, which he himself at one time accepted, is a theory of the development and working of the mind which was evolved by a Jew who has studied chiefly Jewish patients; and it seems to appeal strongly to Jews; many, perhaps the majority, of those physicians who accept it as a new gospel, a new revelation, are Jews. It looks

as though this theory, which to me and to most men of my sort seems so strange, bizarre, and fantastic, may be approximately true of the Jewish race.[5]

Among the men of "his sort" that McDougall had in mind here was undoubtedly Maurice Nicoll. The two men shared a Harley Street office with Hugh Crichton-Miller and served together on the staff of the Empire Hospital for Officers during the war. McDougall quoted Nicoll extensively in his own psychological writings while Nicoll echoed his sentiments in a letter to his father:

> But I believe that in our work lies the germ of something very wonderful and it is strange to think it is traceable to Freud—though, as you know, Freud is one thing and his American and Jewish followers another, for they are all Jews, and it is a kind of Jewish revival of thought—a sort of archaism—from which all the Christians who were entangled in it as I was, have broken free—but not empty-handed.[6]

Jung's first postwar visit to the United Kingdom in 1919 was due to his reputation as one of the leading psychiatrists in Europe and the fact that he was still considered to be a proponent of psychoanalysis. In July he delivered papers to a variety of professional organizations, including the Society of Psychical Research and the Psychiatry Section of the Royal Society of Medicine. The invitation from the latter came from its president William McDougall who took a strong liking to Jung, a feeling that was reciprocated and led to Jung's analyzing McDougall's dreams. McDougall wrote,

> that I have put myself into the hands of Doctor Jung and asked him to explore the depths of my mind, my "collective unconscious"...I have assiduously studied my own dreams under his direction and with his help...I seem to find in myself traces or indications of Doctor Jung's "archetypes" but faint and doubtful traces. Perhaps it is that I am too mongrel-bred to have clear-cut archetypes...[7]

William McDougall (1871–1938) was one of the most famous psychologists of the time. After completing his medical training he went to Borneo to do anthropological fieldwork. He next taught at Oxford and published *An Introduction to Social Psychology* (1908) in which he proposed a theory of human behavior based on animal instincts modified by conscious purpose. He called this the hormic system (from the Greek word *horme*, "purposeful activity").[8] He went to Harvard in 1920 where he became a leading critic of behaviorism. He moved to Duke University in 1927 where he was instrumental in establishing a parapsychology laboratory where he supported the research of J.B. Rhine who was brought from Chicago to join the faculty.

He and Jung had a number of interests in common. The first was a shared dissatisfaction with the scientific materialism of the time. They both felt that the fixation with the experimental method had caused psychologists to overvalue mechanistic explanations of human behavior while dismissing such things as the paranormal as unworthy of study.[9] Both men were influenced by the broadly humane concerns of William James that included extensive investigations of mediums. McDougall's interest was in the mind's innate complexity and organization and felt that Jung's theory of the archetypes would support his lifelong adherence to the Lamarckian theory of the inheritance of acquired characteristics. In one of his last books, McDougall voiced his frustration regarding Jung,

> Hence, although my own experiment on the Lamarckian question has brought me year by year increasingly positive results, my anticipation of the establishment of the archetypes has grown fainter and fainter. Meanwhile Jung has withdrawn himself more and more completely from contact and discussion with common mortals like myself. And the pronouncements which reach this world from the cloud-capped Olympus on which he dwells may have been well calculated to sustain his old converts to the faith, but hardly of a nature to bring new ones into the field.[10]

McDougall's departure from Harvard was not just due to the departmental politics prompted by his vociferous opposition to behaviorism but to the controversy caused by the publication of his book *Is America Safe for Democracy?* in 1921. In it McDougall extolled the innate superiority of the Nordic race for which he has been called "the most indefatigable of the race theorizers among the psychologists of the time."[11] Since the turn of the century racialism had gained a wider hearing in academia and the popular press. In 1899, William Ripley's *The Races of Europe* proposed a division of Europeans into three races: the Nordic, the Alpine, and the Mediterranean. This division involved more than such physical criteria as hair color and skull size but was extended to mental and moral differences. It was picked up by Madison Grant who made it the basis of his 1916 book *The Passing of the Great Race*. America's entry into World War I led to the widespread use of army intelligence tests to evaluate thousands of draftees. The data was misused to lend scientific support to the contention that intelligence was due to inheritance rather than environment.

After the war, the widespread anxieties of Anglo-Saxon Americans found several outlets. For some it meant joining the Ku Klux Klan, which spread from its home base in the South to the Midwest and expanded it list of targets to include Catholics and Jews as well as Negroes. [The D.W. Griffith film *Birth of a Nation* (1915) is credited with inspiring the Klan's revival. Woodrow Wilson had the film screened at the White House and was reported to have said that it was "history written with lightening."] Others like Grant and McDougall saw their role as not merely informing people about this threat to America but as advocating concrete measures to reverse it. This took the form of their support for immigration restriction, which became law in 1924. A quota system that drastically reduced the number of immigrants from Eastern and Southern Europe (homes of the Alpine and Mediterranean "races") was established and remained in effect until 1965. Lothrop Stoddard helped sway public opinion with a series of books and articles about the threats to the Nordic race that became so popular that he rated a reference in *The Great Gatsby.*

Since his first visit to the United States in 1909, Jung was fascinated by the psychological aspects of America's unique racial history. In January 1925 Esther Harding wrote in her diary about a visit by Jung to New York several weeks after his visit to the Pueblo Indians in New Mexico. "[Dr. Jung] spoke on racial psychology and said many interesting things about the ancestors, how they seem to be in the land. As evidence, he spoke about the morphological changes in the skulls of people here in the U.S.A. and in Australia."[12] An extended discussion of the issues referred to here will come later; for now, let it serve as a specific example of how Jung participated in the racialist preoccupations of the time.

Jung contributed the lead article "Your Negroid and Indian Behavior" to the April 1930 issue of the American "magazine of controversy" *Forum* (other articles included "The Dance of Death, Mata Hari's Trial and Execution," "These Women! Dark Reflections on the Fair Sex," and "Prohibition Ten Years Later"). It is important to realize that Jung wrote many articles during this period for the popular press in an effort to disseminate his work to a wider public. Dispensing with technical jargon, Jung wrote with real journalistic flair, presenting himself as a European psychologist with a familiarity with America and Americans. Many of his insights were remarkably perceptive for the time; much of the article deals with the unconscious influence of the Negro upon such white social behaviors as laughter, slang, nonchalance, and appetite for boundless publicity. He also noted the impact of Negro music and dance on American culture. "Incidentally, the rhythm of jazz is the same as the *n'goma*—the African dance. To an accompaniment of jazz music you can dance the *n'goma* perfectly, with all its jumping and rocking and its swinging of shoulders and hips. American music is most obviously pervaded by the African rhythm and the African melody."[13] It should be remembered that Jung knew what he was talking about since he had danced the *n'goma* while visiting East Africa several years earlier.[14]

These perceptive comments are, however, offset by Jung's reliance on such racial stereotypes as the supposed "childlikeness" of the Negro. He also wrote about the danger posed to whites from their continual exposure to blacks, a phenomenon then known as "going black."

> The inferior man exercises a tremendous pull upon civilized beings who are forced to live with him, because he fascinates the inferior layer of our

> psyche... To our unconscious mind contact with primitives recalls not only our childhood, but also our prehistory, and with the Germanic races this means a harking back of only twelve hundred years. The barbarous man in us is still wonderfully strong and he easily yields to the lure of his youthful memories. Therefore he needs very definite defenses. The Latin peoples, being older, don't need to be so much on their guard, hence their attitude toward the Negro is different from that of the Nordics.[15]

One irony here is that Jung soft-pedaled the legacy of slavery and colonialism with the result that it was whites who were "forced" to live with inferior peoples. Jung's racialist argument was given a popular spin by the use of the word "Nordic," which was apparently inserted by an editor since the word is not found in the original written manuscript.

The German Situation

During World War I Allied intellectuals had sought to understand Germany's motives through an investigation of its philosophical tradition. In *Egotism in German Philosophy* George Santayana wrote,

> the Germans have been groping for four hundred years toward a restoration of their primitive heathenism. Germany under the long tutelage of Rome had been like a spirited and poetic child brought up by very old and very worldly foster-parents... it was this elite that made the Reformation, and carried [speculative power and earnestness] on into historical criticism and transcendental philosophy, until in the nineteenth century, in Schopenhauer, Wagner, and Nietzsche, the last remnants of Christian education were discarded and the spontaneous heathen morality of the race reasserted itself in its purity.[16]

After the Germans destroyed the library of Louvain and bombed Rheims Cathedral, references to Nietzsche's "blond beast" multiplied; Allied propagandists cheerfully characterized Germans as modern "barbarians" who were atavistically following the dictates of their Teutonic Volk-Soul.[17] Many Germans decided to take this characterization as a compliment, as an authentic expression of their heathen heritage, and sought to find ways to further it.

In Germany the sharpened cultural divisions caused by the war had led to a split in the Kant Society. In 1916 Bruno Bauch published "On the Concept of the Nation" in the Society's journal, which distinguished between Germans and Jews who were identified as an "alien people." In a follow-up letter he spoke of the need for each group to acknowledge their "folkish difference and destiny." Bauch left the Society and founded the German Philosophical Society whose members were to include Max Wundt, Hans Freyer, Felix

Krueger, and Erich Rothacker.[18] It was this conservative discourse that Jung was joining with his 1918 article "The Role of the Unconscious." Fritz Ringer's analysis of Germany's university mandarins also applies to the non-university intellectuals we will meet:

> It would be wrong to trace the intellectual concerns they shared solely to the theoretical or philosophical antecedents which they had in common. No matter how many German intellectuals of the Weimar period read Kant or Hegel, their manner of thought was not just the product of an inherited logic. It was a certain constellation of attitudes and emotions which united them, infecting even their language and their methods of argument. We must seek to account for the mood which gripped them, not just for their scholarship...[19]

Jung's writings of the 1920s were devoted more to broadly cultural subjects than to clinical ones. Among the constellation of themes that Jung shared with these conservative critics was a generally critical attitude toward the legacy of the Enlightenment. The dichotomy of *Kultur/Civilization* had become a polemical reference point during the war and was carried over into the *Kulturkampf* of the Weimar period. The following schematic list might be the best way to organize all this:

Kultur	***Civilization***
mythos—soul	logos—intellect
spiritual	materialistic
holistic	atomistic
national	international
rural	urban
aristocratic elite	mass democracy
clean—healthy	dirty—degenerate
youthful	senile
life-promoting	hostile-to-life

Scholars of this movement note that although anti-Semitism permeated this discourse, the conservatives' view of the Jews was essentially a traditional, cultural one rather than identical with the "scientific" anti-Semitism first promoted by the turn-of-the-century racial hygiene movement. Most importantly for these thinkers Jews were identified with "modernity" and were held responsible for all the negative effects of that process.

In a 1923 letter Jung developed his analysis of the religious situation of the German people that he began in "The Role of the Unconscious."

> The Germanic tribes [actually, *Rasse*—"race"] when they collided only the day before yesterday with Roman Christianity, were still in the initial state of

polydemonism with polytheistic buds. There was as yet no priesthood and no proper ritual. Like Wotan's oaks, the gods were felled and a wholly incongruous Christianity, born of monotheism on a much higher cultural level, was grafted upon the stumps. The Germanic man is still suffering from this mutilation. I have good reasons for thinking that every step beyond the existing situation has to begin down there among the truncated nature-demons. In other words, there is a whole lot of primitivity in us to make good.

It therefore seems to me a grave error if we graft yet another foreign growth onto our already mutilated condition. It would only make the original injury worse. This craving for things foreign and faraway is a morbid sign. Also, we cannot possibly get beyond our present level of culture unless we receive a powerful impetus from our primitive roots...I find myself obliged to take the opposite road from the one you appear to be following in Darmstadt. It seems to me that you are building high up aloft, erecting an edifice on top of the existing one. But the existing one is rotten. We need some new foundations. We must dig down to the primitive in us, for only out of the conflict between civilized man and the Germanic barbarian will there come what we need: a new experience of God...Shouldn't we rather let God himself speak in spite of our only too comprehensible fear of the primordial experience? I consider it my task and duty to educate my patients and pupils to the point where they can accept the direct demand that is made upon them from within. This path is so difficult that I cannot see how the indispensible sufferings along the way could be supplanted by any kind of technical procedure. Through my study of the early Christian writings I have gained a deep and indelible impression of how dreadfully serious an experience of god is. It will be no different today.[20]

Jung is still concerned about what technique is appropriate for healing the split experienced by Germanic man. What needs to be made clear here is that he was specifically concerned about the applicability of yoga to Europeans since Schmitz had sent him a copy of his book *Psychoanalysis and Yoga* for comment. He felt that yoga, as he had previously felt about psychoanalysis, was not the answer. Each was the product of the unique psychological development of a foreign people so that the answer could only be found closer to home, in the encounter with the primitive man alive in the unconscious. He again referred to the historical precedent for this situation, the traumatic imposition of "Roman" Christianity on the German "race." He explicitly referred to it as "incongruous" and recommended that the encounter with the primitive man within would result in a new experience of God.

The idea that Christianity had created a split in the soul of the German barbarians, a split that was still affecting modern Germans, was not an observation unique to Jung. He was using an idea that had been popular in Germany since the mid-nineteenth century by various writers who sought to foster a coherent identity for the newly emergent nation. One of their main tactics was to emphasize Germany's unique spiritual heritage by tracing it back through

Luther to such German mystics as Meister Eckhart. One of the best-known examples was Richard Wagner's opera *Parsifal* that popularized the Aryan cult of blood purity. One commentator on German life wrote,

> It would be hard to overestimate the significance of this stream of thought in the cultural life of modern Germany. As a religious movement, it took a position consciously opposed to or outside Christian orthodoxy. It was very definitely anti-clerical and in many phases radically anti-Christian... In its claim to express the national religion of the Germans it recognizes folk and folkhood (Volkstum) as something superior to any universal church, such as Christianity, Protestant or Catholic, represents. If it speaks of God, He is thought of as a God of race, a being who stands in a very special relationship to His people, and is conceived of primarily as a Germanic, or Aryan, God.[21]

Roman Christianity and behind it Judaism were seen as alien intrusions that had interrupted the religious development of the German people.

It is clear that the war and the difficult years that followed it provoked Jung to think in far more racialized categories than ever before. But it is also clear that he is harkening back here to the writings of Eduard von Hartmann about religion that he read in his youth. In his 1874 *The Self-Destruction of Christianity and the Religion of the Future* Hartmann compared the differences between the Volk-souls of the Aryan Hindus and the Semitic Jews and Arabs. He felt that the abstract monotheism of the latter had a negative impact on the polytheistic monism of the Aryan peoples. Jung, as we have seen, had already used the work of Hartmann and his follower Drews in the prewar era; as Jung's extensive notations show, he continued to read Drews' work right up until the time of that writer's death in 1935 (Jung's library contained his *Three Essays Bound Together* [Volume C 73]). As we shall see, Jung's would make his sympathy for these Free Church critics of liberal Protestantism clear in his essay *Wotan* (1936).

Jung's contention that the split in Germanic Man would be healed by a primordial experience of the divine shows the impact that Rudolph Otto's *The Idea of the Holy* (1917) had upon him. It explores the phenomenology of numinosity, the experience of awe and dread that is felt in the encounter with "the Other." For Jung this not just a matter of detached observation but was also a deeply personal experience. This can be seen in his account of a dream he had of his mother's death in 1923 around the time of the letter to Schmitz.

> The night before her death I had a frightening dream. I was in a dense, gloomy forest... Suddenly I heard a piercing whistle, and a gigantic wolfhound with a fearful, gaping maw burst forth. At the sight of it the blood froze in my veins. It

> tore past me, and I suddenly knew: the Wild Huntsman had commanded it to carry away a human soul. I awoke in terror...
>
> Seldom had a dream so shaken me...the Wild Huntsman...was Wotan, the god of my Alemannic forefathers, who had gathered my mother to her ancestors...[22]

The recipient of the letter Oscar A.H. Schmitz had became a devotee of Jung's after reading *Psychological Types* and becoming his most enthusiastic promoter in Weimar Germany, writing dozens of articles on Jungian psychology for a variety of newspapers and journals (including the *Zeitschrift für Menschenkunde* started in 1925 in conjunction with Klages' *Journal of Graphology*). Schmitz was also very involved in the School of Wisdom founded in Darmstadt, Germany, by Count Hermann Keyserling in 1919. With Schmitz as his intermediary, Jung also got more involved with the count and his school, eventually lecturing at its annual conference in 1927 and meeting a number of people who influenced his career, the best-known being Richard Wilhelm, the Sinologist who deepened his understanding of Chinese culture.

Before we begin to explore how Jung connected with this new German network, an overview of the general situation of postwar Germany is in order. In the aftermath of its traumatic defeat in 1918, Germany underwent in quick succession the replacement of its monarchy with a parliamentary democracy and the imposition of the humiliating Treaty of Versailles. The trauma continued until 1923 with the country experiencing inflation, political assassinations, several putsch attempts (including one by an obscure Bavarian politician named Adolf Hitler), and the French occupation of the Ruhr, the country's industrial center. The appointment of Gustav Stresemann of the German People's Party to the chancellorship late that year was the turning point. First as chancellor and then as foreign minister he pursued policies that stabilized the country economically and politically. His renegotiation of the reparations payments was followed by a prosperity that was fueled by the infusion of foreign capital. The Locarno Pact of 1925 normalized Germany's relations with its neighbors while its admission to the League of Nations signaled the end of its pariah status. The political scene experienced a welcome respite and the vibrant cultural life that characterized the Weimar Republic was in full swing.

Schmitz conveyed the intellectual flavor of the times in his autobiography *Ergo Sum* (which he dedicated to Jung) in a list of topics fashionable among the international clientele of the great hotels of Europe: cocaine, mahatmas, Richard Strauss, Freud, aspects of Saturn, Max Reinhardt, Dadaism, birth control, Dostoevsky, Ford, Tao, phonographs, spirit-photography, jazz,

Einstein, Dionysos, sadism, Picasso, homosexuality, and yogurt.[23] It is clear that Schmitz was around when the 1920s began to roar!

It is likely that it was Schmitz who introduced Jung to the work of Bruno Goetz, an acquaintance from their days in the Stefan George Circle. In "Wotan" Jung wrote

> In his *Reich ohne Raum* [*Empire without Space*], first published in 1919, Bruno Goetz saw the secret of coming events in Germany in the form of a very strange vision. I have never forgotten this little book, for it struck me at the time as a forecast of the German weather. It anticipates the conflict between the realm of ideas and life, between Wotan's dual nature as a god of storm and a god of secret musings. Wotan disappeared when his oaks fell and appeared again when the Christian God proved too weak to save Christendom from fratricidal slaughter. When the Holy Father in Rome could only impotently lament before God the fate of *grex segregatus,* the one-eyed old hunter, on the edges of the German forest, laughed and saddled Sleipnir.[24]

The novel is the story of Melchior von Lindenhuis who comes between two rival figures: one of whom, Fo, travels with a troop of boys who trigger destructive outbursts in every place they visit. Fo is a nature spirit who proclaims a message of ecstatic surrender to the eternal rhythms of life and death. His adversary is Ulrich von Spat, the ruler of a group of "glass lords" who offer an alternative philosophy to mankind, which is based on order, ethics, and pure spirituality. Melchior has an ambivalent relationship to von Spat but in the end kills his shadow and then dies, united in death with Fo.

This synopsis is taken from an article by Marie Louise von Franz who uses the book as an illustration of the negative aspect of the puer (child) archetype.[25] An archetypal approach dehistoricizes a work by emphasizing its mythological dimension, imagining it to be the product of forces that transcend the personal experience of its creator. The role of the unconscious takes precedence over the author's conscious literary intentions. Although Jung's partiality to "visionary literature" led him to downplay the novel's allegorical subtext, both von Franz and Jung were aware that the book had a connection to events in Germany, namely the rise of Nazism. It cannot be understood without a knowledge of the German youth movement, which became more politically active after the war.[26] Von Franz mentions the poet Stefan George and his erstwhile follower and rival, Ludwig Klages. The novel should, in fact, be read as an allegory of the rivalry between these two men.

Allegory was a popular genre in Germany at the time. Examples include Herman Hesse's *Journey to the East* and Franziska ("Fanny") Countess zu Reventlow's novel *Mr. Dame's Notebooks*, which was a thinly fictional account of her involvement in the Schwabing scene. The climax of the novel comes at a masquerade party where a feud breaks out between two factions of the

"Enormous Folk," one led by the master (Stefan George) and the other by Hallwig (Klages). The main cause of the dispute was Hallwig's anti-Semitism. This is historically accurate since Klages left the George Circle because of the number of Jews George had admitted to it. Like their counterparts in *Reich ohne Raum*, George advocated an austere philosophy of spiritualized aestheticism while Klages championed a philosophy of life based on blood and the instincts (Klages characterized this as the conflict between the "logocentric" and the "biocentric" points of view).

Keyserling and the School of Wisdom

Besides getting to know Jung, Schmitz had also become an intimate of Count Hermann Keyserling (1880–1946), a Baltic German who later left Russia and settled in Germany, marrying a granddaughter of Bismarck. His trip around the world before World War I inspired his international best-seller *Travel Diaries of a Philosopher* (1919). With the patronage of Grand Duke Ernst Ludwig of Hesse he founded the School of Wisdom in Darmstadt, which lasted until the early 1930s. The Grand Duke was a well-known patron of the avant-garde having sponsored an artists' colony in Darmstadt early in the century that had drawn to it such up-and-coming talents as Peter Behrens, a founding father of modern architecture.

Now virtually forgotten, Keyserling was an important figure in the intellectual life of Weimar Germany. His eclectic mix of Eastern spirituality, observations on the "psychology of nations" with a dash of Jungian psychology added later proved appealing to Germans looking for a philosophy that would help them find meaning in a new, postwar world. The School sponsored a series of courses and annual conferences as well as publishing books, a newsletter, and a journal *Der Leuchter* (*The Candelabra*) in association with Otto Reichl Verlag of Darmstadt.

Keyserling saw his school as a training ground for a spiritual aristocracy that would help create a new European culture.[27] Richard Noll's assertion that "he was unabashedly a völkisch German in his metaphysical outlook"[28] is, like so many of his statements, grossly inaccurate and yet another reason to consider his conclusions skeptically. As we shall see, Noll does pick up on Keyserling's racialist vocabulary but then forces Keyserling onto a Procrustean bed that mutilates the rest of his philosophy. An analysis of the contributors to Keyserling's publications and the presenters at his conferences do not bear out Noll's claim. A good example is Keyserling's *The Book of Marriage* (1925). The changing relations between the sexes were a hot topic at the time and Keyserling was right there with an anthology. With his contacts he was able to enlist an impressive list of contributors that included Leo Frobenius (anthropology), Rabindranath Tagore (Bengali poet and Nobel Prize winner), Richard Wilhelm, Ricarda Huch (novelist), Beatrice Hinkle, Thomas Mann, Ernst Kretschmer (psychiatrist), C.G. Jung, Alfred Adler, Havelock Ellis, and Leo Baeck (chief rabbi of Berlin).

This diverse group defies a single characterization but the Germans among them with whom Jung became most familiar were a part of a movement that can best be described as the "conservative avant-garde." Abreast of the latest developments and troubled by Germany's headlong rush into modernity they found Keyserling and his School a congenial point of gravitation. Like Keyserling and Jung they too had been attracted to the *Lebensphilosophie* movement inspired by the life and work of Friedrich Nietzsche who had suffered a breakdown in 1889. A cult developed around him that appealed to those who found the "Dionysian" Nietzsche inspiring. Alfred Schuler, Ludwig Klages' eccentric fellow-Cosmic, was denied a request to heal Nietzsche's madness using rituals that he adapted from those of pagan Rome. Ivan Belyi, Emil Medtner's fellow Symbolist, suffered a temporary breakdown while visiting Nietzsche's grave in Weimar.

Jung and Keyserling were in frequent contact from the mid-1920s through the early 1930s. They referred to each other in their writings and Jung reviewed several of Keyserling's books. Their involvement was practical as well as intellectual since they both had Harcourt, Brace as their American publisher. After Keyserling made a lecture tour through the United States in 1928 he switched to Harper and Co., which brought out his book *America Set Free* (1929). The magazine *Forum* that carried Jung's 1930 article regularly featured Keyserling. It was through a contact of Keyserling's—Victoria Ocampo, an Argentine literary figure—that Jung's *Psychological Types* appeared in Spanish in 1934. Keyserling had met her on a trip that resulted in his book *South American Meditations* (1932) and introduced her to Jung.[29]

The area of their greatest mutual interest was in the "psychology of nations," a form of psychology that had first been given academic respectability by Wilhelm Wundt's research into *Völkerpsychologie.* As its popularity increased later in the century in the wake of imperialism abroad and nationalistic rivalries on the continent, it became the domain of popularizers who gave it a more racialist slant. While studying in Germany as a young man Keyserling was influenced by his meeting Houston Stewart Chamberlain.[30] Keyserling was also influenced in this line of thinking by his friend Gustave Le Bon, the French writer who wrote the pioneering study of group psychology *The Crowd* (1895) but who also wrote about the psychology of nations and authored anti-Semitic writings that warned of the threat that Jews posed to France.[31]

Keyserling's two books *Europe* and *America Set Free* are filled with his personal observations about the psychology of nations and races and it is obvious that he was quite familiar with the racialist writings of the time. He made frequent references to "Nordics," which, as we have seen, had become an increasingly popular term in the 1920s. Keyserling often uses such Jungian terms as "collective unconscious" and "psychological types" to explain himself. That their influence was mutual can be seen in one revealing example. In *Europe*, Keyserling writes that "the old Roman type originally had a substratum of Nordic blood, like the Lombards of today."[32] This idea had originally come from Chamberlain.[33] It was passed on to Jung who later quoted it almost verbatim in his *Zarathustra Seminars* where he said "Fascism in Italy is old Wotan again; it is all Germanic blood down there with no trace of the Romans; they are Langobards, and they have that Germanic spirit."[34] In his 1929 book *The Recovery of Truth*, Keyserling rephrases a statement that was partially cited earlier

> Jung thinks that Freud's presuppositions often apply to Jews and much more rarely to the Nordic type. He holds the characteristics of the unconscious to be dependent on the history of the races, on their ages and destinies; according to him, the Nordic's unconscious is on the whole barbaric and primitive, and correspondingly, unerotic, whereas the Jew with his far-reaching historic past is, within that same strata, a differentiated Alexandrian.[35]

The only change is the substitution of the newly popular "Nordic" for the previously used "Germanic."

It is important to note that for all his use of racialist vocabulary, Keyserling did not subscribe to the kind of biological racism that had become a cornerstone of Nazi ideology. "In Germany anyone who places the accent mark on blood rather than spirit is in the deepest sense of the word a racial alien and not the person in whose veins Nordic blood flows."[36] Keyserling saw *race* as just one factor in the formulation of a philosophy of humanity that also included the *spirit* and the *environment*. Keyserling was to be harassed by the Nazis after their take-over, which he characterized as "the rule of the lower middle class and the dictatorship of the non-intellectuals. Artists, authors, intellectuals of every kind have ceased to be of importance."[37] This was to an end to Keyserling's plans for the formation of a new cultural elite at his School of Wisdom.

The most important of the annual conferences of the School was that held in 1927 with the theme "Man and Earth." Among the presenters were Jung, Richard Wilhelm, the anthropologist Leo Frobenius, the philosopher Max Scheler, and Hans Prinzhorn, a psychiatrist and colleague of Ludwig Klages. Although not in attendance Klages was present in spirit since the theme had been adopted from the title of a book he had written in 1913. The majority of attendees would have known this and been familiar with his dichotomization of systems of thought into "logocentric" and "biocentric." Prinzhorn had done pioneering work on the art of the insane and made common cause with Klages in his hostility to a mechanized, positivistic model of human nature. Jung found this appealing and later became a contributing editor to the journal *Character and Personality* that Prinzhorn and Jung's old friend William McDougall founded in 1932 and published simultaneously in German and English (Jung's contribution to the second issue was "Sigmund Freud in his Historical Setting," CW 15).

The theme created other resonances since *landscape* had been a German intellectual concern since the time of such Romantic painters as Casper David Friedrich and Carl Gustav Carus. By the mid-1920s there was a Carus renaissance that emanated from the circle around Klages who had earlier revived the reputation of Jacob Bachofen (1815–1887), the Swiss legal historian and mythologist best known for his theory of matriarchy. Klages found Bachofen's interest in the "telluric" ("earthly") forces operative in ancient mythology captivating and made it the basis of his "biocentric" position. In 1925 Klages and C.A.Bernoulli brought out a new edition of Bachofen's 1859 book on grave symbolism. The year before, Bernoulli had published two books on Bachofen, one linking him to Klages and his approach to the study of character.[38]

Jung came at the conference topic from two different angles. First, he had been influenced by fellow presenter Richard Wilhelm's work in Chinese philosophy, particularly his research into the life and work of Lao Tzu and the Taoist school. Jung had experimented with the *I Ching*, which was translated into German by Wilhelm and the two men were soon to collaborate on *The Secret of the Golden Flower* (1929), a work of Chinese alchemy. At the time of the conference Jung was preoccupied with relating the ideograms "Yin" and "Yang" to his archetypes of the anima and the shadow. Yin relates to the dark, feminine powers of the earth and something of this sensibility is expressed in the original title of Jung's lecture "Der Erdbedingtheit der Psyche" ("The Conditioning of the Psyche by the Earth"). It first appeared in the 1927 issue of *Der Leuchter* (English translation "Mind and Earth" in *Contributions to Analytical Psychology* [1928]) but was soon divided into two articles that appear separately in the Collected Works ("The Structure of the Psyche" [Vol. 8] and "Mind and Earth" [Vol. 10]). The earth/spirit dyad was one of the defining themes in Jung's intellectual relationship with Keyserling. But, as Jung's criticism of "Darmstadt" in his letter to Schmitz suggested, the Swiss psychoanalyst found the count's spiritual interests philosophically vague and rather pretentious; he recommended, instead, focusing on the primitive (telluric) elements in the Germanic unconscious.

This leads to the second approach that Jung found important. He wanted to supplement his interest in Germany's barbarian prehistory with what he had learned from his recent field trips to the Pueblo Indians of the American Southwest and the Elgonyi tribe of East Africa. They were tribal cultures that had been able to maintain their mythological integrity in spite of the inroads of Western imperialism. For men like Jung and Frobenius they represented the last living examples of a symbolic mentality that characterized all members of the human race before the rise of civilization. They both felt that an appreciation for humankind's mythological heritage would help overcome the spiritual malaise facing modern civilization.

Most of Jung's long paper was an exposition of his theory of the collective unconscious and relied on his usual mix of cultural and clinical examples. At the point in the paper where it was later divided Jung described the archetypes as

> essentially the chthonic portion of the mind—if we may use this expression—that portion through which the mind is linked to nature, or in which, at least, its relatedness to the earth and the universe seems most comprehensible. In these primordial images the effect of the earth and its laws upon the mind is clearest to us...[39]

He continued his argument with his thoughts about "night religion," *participation mystique*, and the anima.

Jung concluded his paper by relating the theme of the conditioning of the psyche by the earth to his experiences with Americans and their country. Jung had visited the country four times before his 1925 visit to the American Southwest. He had learned English early in his career and attracted a sizable American clientele, a fact that elicited comments like this from William McDougall "I have heard rumours to the effect that Dr. Jung, in the intervals between curing various millionaire American neurotics, was making expeditions to study the dreams of various primitive peoples."[40] Jung characterized the typical American as "A European with negro manners and an Indian soul!"[41] The anecdote that he then used to illustrate the mysterious relationship between people and their land was based on an experience he had in Buffalo on his 1909 trip to the United States. He made a passing reference to it in his "The Role of the Unconscious" and used it again in the *Forum* article. He had stood outside a factory door in Buffalo watching the workers exit; he commented to his American companion A.A. Brill that he was surprised by the high percentage of Indian blood that he noticed. When Brill disagreed, Jung concluded that if hereditary did not explain what he had seen, it could be explained by the "mysterious Indianization of the American people." He proposed that Europeans settlers to the North American continent (and Australia as well) were subject to the psychic imprint of the natives of the land. To support this conclusion he made use of a recently released government report on immigration conducted by the German-born anthropologist Franz Boas. What Jung found of most interest was evidence of significant changes in the bodily form, especially the shape of the skull, which had been considered a reliable index of race since it was considered a stable anatomical feature. This landmark field study concluded that these changes were due to such environmental factors as intermarriage, family size, and nutrition.

It is important to note that there is no mention of "Indianization" in the original report, which was Jung's own creation. Returning to the anecdote, Jung claimed that he had observed a "Yankee type" created by contact with the American earth that was physically conforming to a native "Indian type." This conclusion becomes dubious when we realize that the majority of workers Jung was observing were not native "Yankees" at all but recent Slavic immigrants from Eastern Europe who were then pouring into the factories of Buffalo, Chicago, and other Midwestern cities. Ironically, the "Indian" features that Jung observed could in fact be explained in hereditary terms as being due to the influence of the Mongol invasions of Eastern Europe. Jung had even conveyed some sense of this in his 1925 Seminar where he said "I was enormously struck by the resemblance of the Indian women of the Pueblos to the Swiss women in Canton Appenzell where we have descendants of Mongolian invaders. These might be ways of explaining the fact that something in American psychology leans toward the East."[42]

In trying to describe his working method Jung used a word—*Menschenkenner*—that was comprehensible to his original German-speaking audience, many of them readers of the *Zeitschrift für Menschenkunde*, but for which there is no exact English equivalent, being translated as "student of human nature." This approach is rooted in the scientific writing of Goethe and relied heavily on the use of the intuitive faculty of the mind to perceive patterns that the rational intellect prefers to dissect. It was championed as a unique German contribution to science that refused to accept the ascendancy of the experimental basis of nineteenth-century science. It received popular treatment in such books as Chamberlain's *Goethe* (1912) and *Kant* (1916), both of which Jung owned.[43]

Unfortunately, in this case Jung pursued a dubious line of reasoning, using scientific findings to support idiosyncratic conclusions at odds with those of the researcher. Putting a premium on his own talent for *Menschenkenntnis* Jung did not realize the extent to which such an approach could be used to rationalize prejudices. This is evident in an example he gave at the beginning of his discussion where he said "At our elbows we can observe in the Jews of the various European countries noticeable differences..."[44] He goes on to list a number of different Jewish types and further distinguished among a variety of different Russian Jews: Polish, North Russian, and Kossack. In his self-assurance Jung did not realize that this was problematic since there was no such thing as a "Kossack type" Jew. The Russian Orthodox Cossacks were notorious anti-Semites and responsible for numerous pogroms in the Ukraine, most recently during the Russian civil war. Even his opening remark plays to the social prejudices of his audience by its suggestion of having to rub elbows with Jews. All this should put one on guard when considering Jung's anecdotal anthropologizing.

The following reminiscence of Jung conveys a vivid impression of his personality at this time.

> He ranged not only throughout the global field of psychiatry, but also over present-day social and political subjects, as well as historical and mythological material. He covered philosophy, psychology, medicine, economics, and folklore. He reported his experiences at home and abroad, his prejudices, and opinions on many diverse subjects. He had opinions about practically everything, and dwelled especially on the personality structure of Americans, Orientals, Teutons, Jews, and Blacks—Goethe, Nietzsche, and especially Heraclitus. Much of his Weltanschauung seems dated when I read it today, but his philosophy, with its roots in myth and history, is timeless.[45]

During the nineteenth century "Characterology" had developed into a recognized subspecialty in the field of German psychology. It gained scientific respectability among academic psychologists with a concern for types that incorporated such work as Kretschmer's on body types, Jaensch's on eidetic types, and Jung's on psychological types. It was particularly popular among graphologists (among whom was Max Pulver who later joined Jung's circle in Zurich) and an eclectic group of psychotherapists that grew in numbers during the 1920s, many of whom would soon join the General Medical Society when it was founded in 1926.

In his address to the 1927 School of Wisdom conference the philosopher Max Scheler characterized many of the speakers present in the following words "[Ludwig Klages] is primarily responsible for providing the philosophical foundations for the pan-romantic conception of man which we now find among many thinkers in different scientific disciplines, for example Edgar Dacque, Leo Frobenius, C.G Jung, H. Prinzhorn, Theodore Lessing, and to a certain extent, Oswald Spengler."[46] Although he incorrectly suggested that Jung derived his ideas from Klages, Scheler was correct in discerning Jung's affinity to a new circle of intellectuals, his first since leaving the psychoanalytic movement. In spite of their many different interests and points of view, all these thinkers did share a concern for the deeper dimensions of the human experience, exploring it with variations on the intuitive, symbolic epistemology pioneered by Goethe. This was accompanied by a "nonpolitical" stance that had a distinctly conservative slant.

By the late 1920s Jung had become involved with a network that was part of Germany's influential neoconservative movement. It is described by Stark this way:

> The notion of neoconservatism has often been used by historians to identify the radical non-Nazi German Right between 1918 and 1933—those "Trotskyites of Nazism" who leveled a scathing critique from the right on Germany's postwar

> liberal, democratic order and called for a "conservative revolution" to overthrow the Weimar Republic, but who at the same time were distinct from, often critical of, and not infrequently persecuted by Hitler's National Socialist movement...German neoconservatism in the late nineteenth and early twentieth centuries was largely a movement of insecure segments of the middle class, especially the cultivated intelligentsia and various marginal petty-bourgeoisie strata who felt threatened by the entire process of modernization; neoconservatism was, in essence, a manifestation of their anti-modern anxieties.[47]

Stark says that what distinguished it from traditional conservatism was its preference for such themes as spiritual freedom and the creative personality over crass economic self-interest. Instead of organizing themselves into traditional political parties "neoconservatives preferred to organize networks of small clubs, associations, societies, and schools, and to work through the published media to preach their absolute ideals and utopian programs."[48]

Jung benefited from the neoconservative publishing boom that took place at this time. Between 1921 and 1944, Jung did not publish any original, new book; those that appeared were either new editions of older works or anthologies of articles that were appearing in various contemporary journals. Niels Kampmann Verlag of Celle, which published the *Zeitschrift für Menschenkunde* and Keyserling's *Book of Marriage* (both of which Jung appeared in), also published books by Klages and brought out Jung's *Analytical Psychology and Education* in 1926. Through Keyserling Jung was introduced to editors of the Otto Reichl Verlag of Darmstadt that had been publishing the count's works.

It was with Reichl rather than his regular Zurich publisher Rascher that Jung brought out in 1928 *The Relationship Between the Ego and the Unconscious*, a greatly expanded version of a work that first appeared in 1916 as *The Structure of the Unconscious*. The most significant addition to this work appears in his discussion of the collective psyche. "[A collective attitude] means a ruthless disregard not only of individual differences but also of differences of a more general kind within the collective psyche itself, as for example differences of race."[49] Jung elaborated on this statement in a footnote where he identified the following races: Aryan, Semitic, Hamitic, and Mongolian. This indicates that Jung had adopted racial categories current in the German-speaking world that were at variance with those generally accepted in the Anglo-American world. To appreciate what Jung was getting at, an accurate translation of the footnote's first sentence is necessary. "Thus it is a quite unpardonable mistake if we accept the conclusions of a Jewish psychology as generally valid!" ("So ist es ein ganz unverzeihlicher Irrtum, wenn wir die Ergebnisse einer jüdischen Psychologie fur allgemeingültig halten!") The English translation found in the versions of 1928 and the Collected Works differs from this in two significant ways: the exclamation point is replaced by a period and the "we" disappears. This manipulation blunts what was intended to be a highly emotional wake-up call to his readers. He went on to say that "with the beginning of racial differentiation essential differences are developed in the collective psyche as well. For this reason we cannot transplant the spirit of a foreign race *in globo* into our own mentality without sensible injury to the latter..." Involvement with other systems is then not merely an intellectual mistake but an invitation to injury as well. As we have seen, Jung had for ten years been articulating in public and in private his conviction that Freudian psychology had to be understood in racial terms. Unfortunately, he failed then and later to realize the emotional investment he had made in this position and so dismissed the allegations of anti-Semitism made against him by his critics as "cheap accusations."

Walter Struve points out that Reichl had stopped publishing Keyserling's books and School of Wisdom literature in 1927. It was apparently more of a dispute over politics than money since the house had begun to publish many pro-Nazi works that Keyserling found unacceptable.[50] It is possible that an editor familiar with Jung's views on racial differences and the threat of Jewish psychology solicited the footnote. In the coming years Jung's polemicizing against Freudian psychology became a staple in his popular articles and public lectures.

Kulturbund Activities

Besides intellectuals such as Wilhelm, Jung met a number of other people through the School of Wisdom who were to influence his career. The most important of them was Prince Karl Anton Rohan (1898–1975), an Austrian aristocrat who was descended from an old French family that had fled France after the French Revolution. He became an early admirer of Italian fascism and actively promoted a neoconservative agenda (he would be incarcerated for two years after World War II for his Nazi sympathies). He promoted this agenda through two outlets: the Transnational Intellectual Union ("Kulturbund" in German), an association of intellectuals grouped in chapters in cities around Europe, which sponsored lecture programs and an annual conference. Something of its flavor is expressed in a passage from a preliminary version of the group's manifesto

> True culture demands not only the creative force of individual ingenuity, but also a social caste, trained by tradition for receiving and promoting the work of the creating mind, and helping to mold it. Therefore we desire to unite all such supporters of tradition as are willing to help in reforming...the problems of petrification and destruction.[51]

In *Ergo Sum* Schmitz, who had met Rohan in 1921, discussed the different wings of the internationalist movement active in Europe. He contrasted the Pan Europa Movement founded by Count Coudenhove-Kalergi on democratic and humanistic principles with Rohan's organization: "At any rate, Rohan is not democratic and not middle class-liberal, and a person can ask whether he is more revolutionary or more conservative."[52]

It seems clear that Jung and Rohan hit it off, likely after meeting at the 1927 School of Wisdom conference, since Jung gave the first half of his lecture there to the Vienna Kulturbund in 1928 (returning in 1931 and 1932). It was here that Jung first met Jolande Jacobi, the branch's secretary who later fled Vienna after the Nazi Anschluss of 1938, moving to Zurich where she became a Jungian analyst and cofounder of the Jung Institute in 1948. These lectures, along with other articles by Jung, were published in the other outlet for Rohan's views, the *Europäische Revue*, a journal he founded in 1925 and edited until 1936. It mixed political, economic, and cultural articles and was identified by Armin Mohler as one of Germany's leading "young conservative" publications.[53] Jung published nine articles in it between 1927 and 1934 (a rate of one every nine months) making it the journal he was most closely associated with during these years. In a newspaper article entitled "The Fight against Neurosis and the Renewal of Europe" Rohan wrote that "Jung stands among the leading avant-garde in the fight for a new Europe."[54] Jung's articles also began to appear regularly in the Spanish journal *Revista De Occidente* of Ortega y Gasset who was active in the Kulturbund and the author of *The Revolt of the Masses*, a classic defense of the aristocratic principle in society.

The most important of Jung's articles to appear in the *Europäische Revue* was "The Spiritual Problem of Modern Man," which appeared in the journal's December 1928 issue. It was the text of the lecture that Jung had recently delivered to the Kulturbund's annual conference held that year in Prague from October 1 to 3. The theme was "Elements of Modern Civilization" and its one hundred fifty delegates included Richard Wilhelm and the architect Le Corbusier who as a Swiss lectured in conjunction with Jung. Other attendees included two German intellectuals, the poet Rudolf Binding and the philosopher Leopold Ziegler, who would both cross Jung's path again.[55]

In this paper Jung developed his most explicit and extended analysis of modernity. He defined modern man as an individual who had chosen to adopt an unhistorical attitude toward the past, choosing to live consciously in the present moment. That Jung's admiration for these individuals is matched by a disdain for what he dubbed the *Auch-Moderne* ("pseudo-moderns") becomes clear in the following passage.

> A great horde of worthless people do in fact give themselves a deceptive air of modernity by skipping the various stages of development and the tasks of life

they represent. Suddenly they appear by the side of the truly modern man—uprooted wraiths, bloodsucking ghosts whose emptiness casts discredit upon him in his unenviable loneliness. Thus it is that the few present-day men are seen by the undiscerning eyes of the masses only through the dismal veil of those spectres, the pseudo-moderns, and are confused with them... This, however, should not prevent us from taking it [proficiency] as our criterion of the modern man. We are even forced to do so, for unless he is proficient, the man who claims to be modern is nothing but a trickster. He must be proficient in the highest degree, for unless he can atone by creative ability for his break with tradition, he is merely disloyal to the past. To deny the past for the sake of the being conscious only of the present would be sheer futility. Today has meaning only if it stands between yesterday and tomorrow. It is a process of transition that forms the link of past and future. Only the man who is conscious of the present in this sense may call himself modern. Many people call themselves modern—especially the pseudo-moderns. Therefore the really modern man is often found among those who call themselves old-fashioned.[56]

If we recall Jung's oft-repeated comment that psychologizing is a "subjective confession," this definition can be understood as an example of his own self-definition and a flattering characterization of his audience. In rhetoric evoking Goethe and Nietzsche Jung tried to unlock the psychological meaning of the contemporary *Zeitgeist*. Dating the revolution in outlook back to the catastrophic results of World War I, Jung wrote that "as for ideals, neither the Christian Church, nor the brotherhood of man, nor international social democracy, nor the solidarity of economic interests has stood up to the acid test of reality."[57] He observed that various cultural trends of the day such as theosophy, anthroposophy, and spiritualism were overtaking Christianity in popularity.

His most sarcastic comments were aimed at psychoanalysis:

Freud... has taken the greatest pains to throw as glaring a light as possible on the dirt and darkness and evil of the psychic background, and to interpret it in such a way as to make us lose all desire to look for anything behind it except refuse and smut. He did not succeed, and his attempt at deterrence has even brought about the exact opposite—an admiration for all this filth.[58]

Then, "men like Havelock Ellis and Freud have dealt with like matters in serious treatises which have been accorded all scientific honours. Their reading public is scattered over the breadth of the civilized, white world. How are we to explain this zeal, this almost fanatical worship of everything unsavoury?"[59] "There are too many persons to whom Freudian psychology is dearer than the Gospels, and to whom Bolshevism means more than civic virtue. And yet they are all our brothers, and in each of us there is at least one voice which seconds them, for in the end there is one psyche which embraces us all."[60]

Finally, No wonder that unearthing the psyche is like undertaking a full-scale drainage operation. Only a great idealist like Freud could devote a lifetime to such unclean work. It was not he who caused the bad smell, but all of us—we who think ourselves clean and decent from sheer ignorance and the grossest self-deception."[61] Jung sided with the "silent folk of the land" who with their spiritual yearnings had a more instinctive sense of developments in the collective psyche than the intellectual celebrities whom he derided for their academic prejudices.

The titles of Jung's 1931 and 1932 Kulturbund lectures—"The Unveiling of the Psyche" ["Die Entschleierung der Seele"] and "The Inner Voice" ["Die Stimme des Innern"]—evoke a Symbolist sensibility familiar to his Viennese audience that is missing from their translations into "The Basic Postulates of Analytical Psychology" (CW 8) and "The Development of Personality" (CW 17). These titles were apparently chosen to convey an impression of Anglo-American pragmatism that muted the original German tone. Much of Jung's argument in the first lecture is his critique of nineteenth-century materialism as embodied in positivistic science,

> if we maintain mental and psychic phenomena arise from the activity of the glands we can be sure of the respect of our contemporaries, whereas if we attempted to explain the break up of atoms in the sun as an emanation of the creative *Weltgeist* we should be looked upon as intellectual cranks. And yet both views are equally logical, equally metaphysical, equally arbitrary and equally symbolic. From the standpoint of epistemology it is just as admissible to derive animals from the human species as man from the animal species. But we know how ill Dacque fared in his academic career because of his sin against the spirit of the age, which will not let itself be trifled with. It is a religion or, better, a creed which has absolutely no connection with reason, but whose significance lies in the unpleasant fact that it is taken as the absolute measure of all truth and is supposed always to have common sense on its side.[62]

Jung's representative martyr to the scientific spirit of the age was an interesting choice. Edgar Dacque (1878–1945) was a German paleo-geologist who had established a respectable reputation in his field before going on to develop an eccentric theory that rejected the Darwinian explanation of evolution. He claimed that the human race had begun millions of years ago and that the various primate families were off-shoots of it. It was an example of an extreme form of metaphysical idealism that emphasized intuition while denying the implications of the slowly growing body of fossil evidence. In its claims it went beyond the idealistic natural philosophy of such early-nineteenth-century figures as Schelling to incorporate a great deal of occult speculation that Dacque acquired through his membership in the Theosophical Society.[63]

His most famous book *Primeval World, Legend, and Humanity* (1924) is a good example of the kind of fringe science that gained respectability among large segments of the German middle class during the 1920s. Other popular theories included Hans Hörbiger's Cosmic Ice Theory and the Atlantis theory of Herman Wirth that proposed a northern locale, Thule, as the site of that legendary kingdom. These all found advocates among the early Nazi leadership. (In 1935 Himmler founded the *Ahnenerbe* ["Ancestral Heritage"] with Herman Wirth in an early leadership position.) Hermann Rauschning recalled

> A savant of Munich [i.e., Dacque], author of some scientific works, had also written some curious stuff about the prehistoric world, about myths and visions of early man, about forms of perception and supernatural powers. There was the eye of the Cyclops or median eye, the organ of magic perception of the Infinite now reduced to a rudimentary pineal gland. Speculations of this sort fascinated Hitler, and he would sometimes be entirely wrapped up in them. He saw his own remarkable career as a confirmation of hidden powers. He saw himself as chosen for superhuman tasks, as the prophet of the rebirth of man in a new form.[64]

It seems likely that Jung became acquainted with the work of Dacque through his friendship with Gustav Richard Heyer. Heyer was a Munich psychotherapist who had helped found the General Medical Society for Psychotherapy in 1926. He was to eclipse Oscar Schmitz as Jung's leading promoter in Germany after founding a Jungian study group that was the nucleus of the German language seminars held by Jung in Küsnacht in October 1930 and 1931. Heyer also introduced Jung to the Indologist Jacob Wilhelm Hauer who conducted a seminar on Kundalini Yoga in Küsnacht in 1932 as well as making a presentation on yoga and psychotherapy to the General Medical Society's conference in the same year.[65] Jung's relationship with Hauer deepened and grew more complicated after Hauer founded the German Faith Movement in 1933. As we shall see, Hauer's perspective was a major influence on how Jung was to interpret developments in Nazi Germany.

Jung was familiar with much of this fringe literature and used some of it in eclectic support of his theory of the collective unconscious. In turn Dacque, who was also being published in the *Europäische Revue*, cited Jung's ideas about mythology in his 1938 *Lost Paradise.* (Via Heyer, Dacque's influence is evident in Jean Gebser's *The Everpresent Past*, which has become popular with New Age intellectuals.) Jung began to list Herman Wirth's *The Ascent of Humanity* (1928) in the bibliography of later editions of *Symbols of Transformation.* Although he doesn't cite Wirth in the text, he does use an illustration from that book as plate Ib in his own. It is identified as a "Sun-god, Shamanistic

Eskimo idol, Alaska." Although Jung never commented on the piece it is interesting to note its place in Wirth's theory.

> Like Churchward, Wirth thought he had the key to the profoundly sacred symbolism of primitive man, so that by tracing the symbols of primitives the world over he could reconstruct the pre-history of man. For instance a pair of circles one above the other, connected by a short line, represents the year. Wirth believed that the last survivors of his arctic civilization were the now extinct Sadlermiut Eskimos, descendants of the Thuleans who flourished between 25,000 and 12,000 B.C., contemporary with the Cro-Magnon men; and that their culture, while high, was non-metallic. They spread to Europe, Asia, and the Americas, splitting into the present racial types as they went, and even migrated as far as New Zealand.[66]

What is of importance here is not so much what Jung believed but what he tolerated intellectually. With his reservations about positivistic science and penchant for symbolic interpretations, he found this literature appealing and failed to give it the critical reading it deserved.

Jung's sensitivity to developments in Germany is evident in the Kulturbund lecture "The Inner Voice" of November 1932. He began with the comment that "the great liberating deeds of world history have always sprung from leading personalities and never from the inert mass."[67] He devotes most of the lecture to the role of the inner voice, the "vocation," in history. "It is not for nothing that our age calls for the redeemer personality, for the one who can emancipate himself from the inescapable grip of the collective and save at least his own soul."[68] The pivotal comment in the lecture is the following "There are times in the world's history—and our own time may be one of them—when the good must stand aside, so that anything destined to be better first appears in evil form."[69] Consciously Jung was evoking the Taoist philosophy familiar to his audience from the work of Richard Wilhelm (as a matter of fact, the concluding sentence of the lecture is "Personality is Tao"). It can also be understood as a rationalization for the conservative acquiescence to the Nazis, a phenomenon that Julien Benda identified in his best-seller as *la trahison de clercs* ("the treason of the intellectuals").

In his lecture, Jung mixed this moral neutrality with a sarcastic critique of the intellect.

> The Age of Enlightenment, which stripped nature and human institutions of gods, overlooked the God of Terror who dwells in the human soul. If anywhere, fear of God is justified in face of the overwhelming supremacy of the psychic. But all this is so much abstraction. Everyone knows that the intellect, the clever jackanapes, can put it this way or any other way he pleases.[70]

In an unpublished book introduction of the same period Jung wrote,

> It is most refreshing, after the whole nineteenth century and a stretch of the twentieth, to see the intellect once more turned loose upon herself...As a matter of fact, it is wholesome and vitalizing tearing into some sorry shreds of what all 'healthy-minded' people believed in as their most cherished securities. I am human enough to enjoy a juicy piece of injustice when it comes in the right moment and in the right place. Sure enough, Intellect has done her worst in our "Western Civilization," and she is still at it with undoubted force.[71]

Like so many at that time, Jung failed to realize that the true threat to Western civilization lay not with the excesses of Reason but with developments in Germany that he was watching with considerable sympathy. He was willing to watch Reason fall from its pedestal hoping that a more symbolically attuned consciousness would take its place.

Jung in the Popular Press

Several days after returning from his lecture in Vienna, Jung's article "Picasso" appeared in the *Neue Zürcher Zeitung* (November 13, 1932). It expressed his personal reactions to the comprehensive Picasso exhibition at the Zurich Kunsthaus that had just closed. His apparent characterization of Picasso's art as "schizophrenic" raised an outcry and led Jung to clarify himself later when the article appeared in an anthology where he made it clear that he was not pathologizing the artist but merely identifying the schizoid nature of Picasso's creative talent. In his review, Jung wrote,

> When I say "he," I mean that personality in Picasso which suffers the underworld fate—the man in him who does not turn towards the day world, but is fatefully drawn into the dark; who follows not the accepted ideals of goodness and beauty, but the demonical attractions of ugliness and evil. It is the antichristian and luciferian forces that well up in the modern man and engender an all-pervading sense of doom veiling the bright world with the mists of Hades, infecting it with deadly decay, and finally, like an earthquake, dissolving it into fragments, fractures, discarded remnants, debris, shreds, and disorganized units. Picasso and his exhibition are a sign of the times, just as much as the twenty-eight thousand people who came to look at his pictures.[72]

Jung had warmed to the role of culture critic but, unfortunately, his insightful comments on various symbolic motifs in Picasso's oeuvre were spoiled by the tone of a diatribe whose sarcasm was more peevish than ironic. Jung associated the modern art sensibility with ugliness and chaos, implying that it was not just a mirror but also a cause of such realities in modern life. His comments reflected the attitudes of the conservative avant-garde who felt ill at ease with the negative tendencies they perceived in such developments as modern art and psychoanalysis.

Besides his Kulturbund activities Jung was also cultivating connections to the Zurich literary scene. In May 1922 he gave a lecture "On the Relation of Analytic Psychology to Poetry" (CW 15) to the local Society for German Language and Literature. In it Jung developed a theory of poetic creation that was inspired by Emil Ermatinger's recently published *Das dichterische Kunstwerk*. Ermatinger (1873–1953) was a professor of literature at the University of Zurich and the ETH where he taught Max Rychner, Max Frisch, and Walter Muschg. He was active in the Lesezirkel Hottingen, the city's leading literary club. Located near the Psychology Club, it invited internationally known intellectuals, including Jung, to lecture. In 1930 Ermatinger published *Philosophie der Literaturwissenschaft* with a contribution from Jung "Psychologie und Dichtung" (CW 15). He would have been a supporter of Jung's reception of the city's Literary Prize in 1932.

Jung's writings of this period drew the attention of Ernst Seilliere, a French critic who wrote a trilogy entitled *Neo-Romanticism in Germany* for the Alcan Press of Paris (1927, 1929, and 1931). Jung owned the first volume where he would have read that Seilliere concurred with Max Scheler's locating him in a neoromantic group of intellectuals that included Prinzhorn and Klages on account of their mutual interest in the symbolic, primitive layer of the human mind. The second volume is devoted to Keyserling and his School of Wisdom; the third discusses Prinzhorn, Klages, Dacque, and Jung; in it Seilliere also discussed the works of Bruno Goetz and Leopold Ziegler. Prinzhorn, author of the *Art of the Insane* and a former student of Bleuler and Jung's at the Burghölzli, had moved squarely into the Klages camp and responded to Seilliere's work with a rejoiner in the spring 1933 issue of the conservative *Deutsche Rundschau* labeling Seilliere a "Romanist" and "royalist," a representative of the French Enlightenment. Extolling the biocentric vision of Klages, he cited a distinguished lineage that included Cusanus, Paracelsus, Goethe, Carus, and Nietzsche. He also castigated Keyserling for what he considered a libelous article about Klages in the January 14, 1933, issue of the *Kölnische Zeitung.*

In it Prinzhorn referred to the newspaper's Swiss feuilleton editor Max Rychner who also was involved in editing the *Neue Schweizer Rundschau*, which had published Jung's *Women in Europe* in book form (1929), a year after it had appeared as an article in *Europäische Revue.* Rychner had drawn close to Rohan and was probably familiar with Jung since his studies at the University of Zurich and certainly from his time as editor of *Wissen und Leben* which published Jung's first article on poetic artwork.

For the *Kölnische Zeitung* Rychner solicited a contribution from Jung in conjunction with the 1932 Goethe centenary. Along with other notables Jung responded to a series of questions about his attitudes about Goethe. To questions about recommending Goethe to the young or to the masses he replied,

> Young people today try to be unhistorical. Goethe does not seem to mean much to them because, for them, he is too close to the fishy ideals of the 19th century. Everything to do with the masses is hateful to me. Anything popularized becomes common. Above all I would not disseminate Goethe, rather cook books.[73]

Jung's connections with men like Rohan and Rychner had led to many publishing opportunities that increased his visibility in the public eye.

Political Affiliations

Although Jung, like most members of the German intelligentsia, always described himself as a "nonpolitical" person, a consideration of his political affiliations and attitudes are of critical importance in understanding his public behavior during the late Weimar years and the subsequent Nazi period. Nonpolitical to the extent that he cared little for traditional party politics and legislation, Jung espoused attitudes that were political to the degree they matched those generally held by the German neoconservatives of the 1920s. Given his new post-Freudian network, it is not surprising that he articulated views similar to theirs and those of the audiences he acquired through them.

As Struve noted, the neoconservatives of the 1920s and early 1930s all shared the goal of cultivating a new elite to lead Germany. What distinguished them from the Nazis, and drew their ire, was the fact that biological racism did not play a part in the neoconservatives' agenda.[74] This would be true of Count Keyserling and of his School of Wisdom associates, whose primary aim was to cultivate a new spiritual aristocracy composed of individuals who had cultivated new levels of self-awareness and so proved their capacity to lead a rejuvenated society. It is important to note that German neoconservatives conceived of "the individual" in a very different way than was common in the Western democracies. They based their concept on ideas developed during the romantic reaction to the French Revolution.

> Romantic individualism must be sharply distinguished from atomistic individualism...[it] stressed the uniqueness of individuals, a uniqueness which placed them beyond conformity to any general law or principle...A

> *Personlichkeit* ["personality"] is one who is distinct, not subordinate, cannot be counted or numbered with others. Goethe was an inspiration for this kind of individualism.[75]

It is from this perspective that Jung's article "The Inner Voice" ("The Development of the Personality") must be read. Opening with a quote from Goethe, Jung writes "Personality is the supreme realization of the innate idiosyncrasies of a living being. It is an act of high courage flung in the face of life, the absolute affirmation of all that constitutes the individual, the most successful adaptation to the universal conditions of existence coupled with the greatest possible freedom for self-determination."[76] This and Jung's other writings on education and personality development reflect the romantic preoccupation with the affirmation of creative self-expression, as well as a patrician disdain for the masses rooted in his Basel upbringing.

What needs to be appreciated here is the extent to which members of Jung's new network mostly belonged to one of Germany's major political parties, the Deutsche Volkspartei (DVP) or German People's Party. The party had been founded in December 1918 and was led by Gustav Stresemann until his untimely death in 1929. It was a conservative, middle-class party that generally represented the interests of big business. Jung's comment that "everything to do with the masses is hateful to me" reads like a direct quote from the 1931 program of the DVP, in which the party deplored "the rule of the masses." Other party concerns included opposition to "exaggerated parliamentarianism" and the need for a new concept of leadership; Jung would express similar views in an interview he would be giving soon to Radio Berlin.[77]

Jung's comment about the "masses" had appeared in the *Kölnische Zeitung* as did Keyserling's critique of Klages; Oscar Schmitz was a regular contributor to this newspaper. The *KZ* was originally an organ of the National Liberal Party and during World War I expressed the views of the German Foreign Office, but had now become a major supporter of the DVP (as was the *Berliner Börsen-Zeitung*, which would publish a laudatory article about Jung in the first months of the Nazi period).[78] The paper ran an ad in the December 1931 issue of the *Neue Schweizer Rundschau* that stated "it was prepared to fight for the spiritual and national freedom and for privately initiated culture and economic arrangements."

Although his affiliation with the DVP is still unclear, Prince Karl Rohan had contacts with the party's conservative wing. In light of his opposition to progressive internationalism, it is not surprising that the head of the German delegation to the 1928 Prague Conference was Baron Kurt von Lersner, the most high profile opponent within the DVP of Stresemann's rapprochement policy.[79] Also, after the onset of the Great Depression, IG-Farben, which was influential within the leadership of the DVP, stepped in to subsidize the

Europäische Revue and other Kulturbund activities. Lilly von Schnitzler, whose husband George was a director of the company, was one of the Kulturbund's treasurers.[80]

Another intellectual with links to the DVP was the philosopher Leopold Ziegler (1881–1959). He had begun his career as a student of Rudolph Eucken and Arthur Drews and published *Western Rationalism and Eros* in 1905. His interest in the role of religion in self-awareness eventually led him to Count Keyserling and his eclectic philosophy of Eastern thought and Jungian psychoanalysis. He lectured at the School of Wisdom and published several books with Reichl Verlag. He coauthored one of them—*The New Aristocracy* (1930)—with Bruno Goetz. In his 1923 autobiography, Ziegler wrote "The outstanding psychologist and philosopher C.G. Jung, one of the few to perceive what religion is, like Buddha speaks with understated courtesy but with firmness when he says 'The question of the existence of God is the stupidest question a man can pose.' "[81] This group shared a strong belief that Europe in general and Germany in particular was in a state of cultural crisis and that the only way out was through spiritual renewal. Where they differed was the form this renewal should take. Ziegler generally supported a new incarnation of the Holy Roman Empire, a society founded on traditional Christian values of authority and order and led by a new spiritual elite.

Ziegler shared this orientation and a personal friendship with one of the major architects of the "conservative revolution" in Germany, Edgar Jung (1894–1934; no relation to Carl Jung). After serving at the front during the war, Jung returned to his native Westphalia where he became a lawyer and active in the local branch of the DVP. He also joined the underground, which opposed the French occupation with violence. He moved to Munich where he published many articles attacking the Weimar parliamentary system. His major critique was *The Rule of the Inferiors,* which was published in an expanded version by the Deutsche Runschau Verlag in 1930. Like many conservatives he recognized the threat posed by the Nazis only after it was too late. He wrote a speech for the conservative politician Franz von Papen who delivered it to the students at Marburg University in June 1934. It contained daring criticisms of the Nazi rule and was immediately suppressed. Jung soon paid for this transgression with his life; he was shot several weeks later during the Night of the Long Knives when Hitler eliminated opponents both inside and outside the Nazi Party.

Carl Jung and Edgar Jung were familiar with each other's work. Edgar Jung mentioned Carl Jung in his *The Meaning of the German Revolution* (1933) where he linked his name with Klages (a common occurrence in the literature of the day) as a modern psychologist who taught a "return to the Mothers."[82] After Edgar Jung's name was mentioned during the June 17,

1936, session of *Zarathustra Seminar*, Jung gave his opinion about the "rule of the inferiors."

> The idea that every man has the same value might be a great metaphysical truth, yet in this space-and-time world it is the most tremendous illusion; nature is thoroughly aristocratic and it is the wildest mistake to assume that every man is equal. That is simply not true. Anyone in his sound senses must know that the mob is just a mob. It is inferior, consisting of inferior types of the human species. If they have immortal souls at all then it is God's business, not ours, we can leave it to him to deal with their immortal souls which are presumably far away, as far away as they are in animals. I am quite inclined to attribute immortal souls to animals; they are just as dignified as the inferior man. That we should deal with the inferior man on our own terms is all wrong. To treat the inferior man as you would treat a superior man is cruel; worse than cruel, it is nonsensical, idiotic.[83]

This blunt statement is arguably the most insensitive statement that Carl Jung ever made. It is, however, not an aberration but entirely consistent with the development of his political and social thinking. The same disdainful Nietzschean tone also permeated the neoconservative discourse in which he was participating. This statement and previous ones by Jung are echoed in Edgar Jung's "Germany and the Conservative Revolution": "in place of equality comes the inner value of the individual," "the fundamental attitude is . . . a religious one," and "the liberal conception of the world has revealed itself as illusory since it has proved impossible to gain mastery over life through abstraction and the rule of understanding."[84]

A final connection Jung had to DVP circles was through Otto Curtius, a Jungian therapist who was to become active in the International General Medical Society for Psychotherapy. His brother Julius had been a DVP Reichstag member from 1920 to 1932 and had served as both finance and foreign ministers.[85] Jung had become active in the General Medical Society after joining it in 1928 and became an honorary vice president in 1930. Among the organization's founders were his old psychoanalytic colleagues Leonard Seif and Hans von Hattingberg. Other founders included Jung's new protégé Gustav Richard Heyer and Carl Häberlin, who headed a sanitorium in Bad Nauheim, which was the site of several of the Society's annual conferences. Häberlin had been an early participant in the School of Wisdom and had an article appear in the 1921/22 issue of *Der Leuchter*. He was a major promoter of the work of Klages, publishing three books about him with Niels Kampfmann Verlag. In the years to come other students of Klages would gravitate to Jung, most notably Martin Ninck whose work on Germanic mythology would have a decisive influence on Jung when he came to write his 1936 article "Wotan."

By 1932 Jung had become very active on the German intellectual scene. He was lecturing to the School of Wisdom, the Kulturbund, and the General Medical Society, seeing the formation of the first Jungian groups, and publishing in a variety of journals. Jung was feeling so confident about this success that he briefly considered starting a journal of his own called *Weltanschauung*. The list of people that Jung considered for the editorial staff and contributions is like a snapshot of his intellectual circle: Heyer as chief editor, then Hauer, Wolfgang Kranefeldt (a young Jungian analyst), Count Keyserling, and Leopold Ziegler.[86] The financial difficulties that killed this venture also helped seal the fate of the Weimar Republic. After the appointment of Hitler to the chancellorship on January 30, 1933, events in Europe and in Jung's career would take a dramatic turn.

Chapter 4

The Question of Accommodation

In the second installment of "On Psychology" (now "The Meaning of Psychology for Modern Man," CW 10, paragraphs 290–312 and 320–321), published in 1933, Jung opened his essay with the following statement:

> We can hardly deny that ours is a time of dissociation and sickness. The political and social conditions, the fragmentation of religion and philosophy, the contending schools of modern art and modern psychology all have one meaning in this respect...The word "crisis," so often heard, is a medical expression which always tells us that the sickness has reached a dangerous climax. (Par. 290)

He then discussed this situation as an analog of that of the Roman Empire in terms that recall Jacob Burckhardt. "Thus, the sickness of dissociation in our world is at the same time a process of recovery, or rather, the climax of a period of pregnancy which heralds the throes of birth. A time of dissociation such as prevailed during the Roman Empire is simultaneously an age of rebirth" (par. 293).

During the early years of the Nazi regime Jung assumed the persona of a physician of society diagnosing its ills with his law of compensation of psychic functioning. Overdevelopment of consciousness leads to dissociation from the unconscious that is then expressed through various symptoms. Although Jung did not refer here to the German situation specifically, his audiences were highly sensitized to a discourse about "crisis" and "regeneration" that was being used by the Nazis to legitimize their claims to power. Public reaction to what was widely perceived as a "conservative-revolutionary" dismantling of the Weimar system was muted. The Nazis quickly began a process called *Gleichschaltung* or "coordination," the forced or voluntary alignment of *all* institutions in Germany to the will of the Führer. Of particular importance in this regard were the takeovers and censorship that quickly became a fact of life in the press and book publishing.

Hans Sluga has written that

> Given the course of German history in our century, it is understandable that those who lived through it were profoundly haunted by a sense of crisis. But in hindsight we can also see that their interpretation of the nature and course of the crisis was generally based on a misjudgment of the historical situation. In retrospect we can see that philosophers and politicians who assumed that they were facing a unique and apocalyptic event from which Germany and the west would emerge deeply changed were deceived about the dynamics of the crisis they were living through. The deception was not based on factual errors. It was due rather to the fact that they were in the grip of a historical a priori. This a priori had led Fichte and Nietzsche earlier to assume that they could discern a unique historical turning point. It subsequently led Heidegger and his contemporaries to postulate such a turn in their lifetime. But they had all been equally deceived.[1]

The views that Jung expressed in his writings of this period clearly reflect just this way of thinking.

Jung had become more a highly visible public intellectual in Germany due to his many lectures, interviews, and involvement in the Kulturbund. He had also been active in the General Medical Society for Psychotherapy, and was to become its president in summer 1933. His decision to accept the presidency and the public statements that he made during his tenure would have both immediate and long-term repercussions for his reputation.

Jung's lecture schedule had grown increasingly busy since his appearance at the School of Wisdom in 1927. Following his December 1932 Kulturbund lecture in Vienna, he embarked on a two-week lecture tour through the Rhineland. Around this time he wrote Smith Ely Jelliffe, his old psychoanalytic colleague, that "My work is unfortunately frequently interrupted by public lectures, particularly in Germany where I seem to be subject to a most inopportune popularity."[2] Jung was back in February to give a Kulturbund-sponsored lecture in Cologne and Essen. The CW incorrectly identifies it as "The Meaning of Psychology for Modern Man," which Jung had delivered at the General Medical Society for Psychotherapy conference in Dresden (1931) and then gave as his acceptance speech for the Zurich Literary Prize (1932). Research shows that the lecture he actually delivered was the Vienna lecture "The Inner Voice," which had been renamed "The Development of Personality."[3]

Jung later related an anecdote from that trip.

> I cannot omit to tell you a little story of what happened to me just two months after the beginning of the revolution. I had a contract to give a course of lectures

in several towns. In the meantime the revolution had come on. There was a big reception at Frankfort [*sic*] [at the Schnitzler's] and many uniformed officers appeared and the situation was a bit uncomfortable, and a man had come down from Berlin—a sort of thief-smeller. He had to smell me out—if my smell was right or wrong. And after dinner while the general conversation was going on, suddenly he said: "Ladies and Gentlemen I want to ask Dr. Jung a question." A deep silence followed—then he said: "Dr. Jung, will you be kind enough to tell us what you think about the swastika?" I thought—now you are in for something, and I said, "You know, as everybody knows it is a very old and widespread symbol, and you can find it all over the world in all sorts of civilizations." The officer said "Oh, we know all that, but you know something more about it, don't you?" I said, "yes, if you want to know, I call your attention to the fact that it turns the wrong way, and it is very unfavorable. That is what people say in the East." Then he looked at me and there was a painful silence—and then a friend of mine made the apt remark, "but you know Doktor, when one is inside the mandala it turns the right way round," and I said "yes, for why are you inside?" Of course they did not understand the joke...After relating that the "thief-smeller" had tried to intimidate him in a rather dissociated way Jung continued "And instinctively he understood what easily happens when you are outside the mandala—then you are happy in an optimistic crowd and you lose the realization of how things really are. You get that feeling only when you are inside the mandala or swastika. Then you are really in the right place."[4]

This anecdote captures the changes in atmosphere and behavior that occurred immediately upon Hitler's ascent to power on January 30, 1933. The new regime was interested in monitoring cultural functions closely and would be taking Jung's opinions seriously. His initial response to the thief-catcher was diplomatic, the implied criticism of the follow-up remark was qualified by attributing it to "what people say in the East." His companion's timely intervention diverted the conversation into harmless wordplay. As Geoffrey Cocks makes clear in *Psychotherapy in the Third Reich* (1997), Nazi Germany would, in fact, provide more opportunities than obstacles to Jung's ambition to promote his psychology.

Jung's Busy German Schedule

Jung's choice of "the inner voice" as the theme for his lectures was timely since it was associated in the minds of his audience with Hitler, whose inner voice was said to have guided him to the German chancellorship. Jung described the leader's sense of being called by his "inner voice" as a paradigm of personality development. When the lecture was first published in the Collected Works it appeared with a footnote that Cocks points out was inaccurately translated as

"After this was written, Germany also turned to a Führer." "The latter verb construction implies a neutrality or even a disparagement on Jung's part and a resignation or desperation on the part of the Germans not expressed by the original language. The translation should read: 'Since this sentence was written, Germany has found its Führer.'"[5] Later editions do include this correct translation. The correct translation makes it clear that Jung saw Hitler's coming to power as the outcome of a natural, almost inevitable process.

Jung's lecture relied on the classical German ideal of *Bildung* ("cultivation") to describe the process of personality-expansion that he called "individuation." He contrasted it to the specialized, textbook approach to education (*Erziehung*) that had come to dominate the field. Jung lectured on "The Development of Personality" to several university audiences that summer and fall. Jung's address at Basel University was reported in the July 4 edition of the city's *National Zeitung*. The reporter parenthetically noted that "Jung polemicized here against his onetime teacher and current adversary Sigmund Freud" when he discussed the infantilizing influence psychoanalysis was having on education. This indicates the way in which Jung would deploy neoconservative concepts of the 1920s to explain his differences with his former psychoanalytic colleagues in the increasingly volatile public context after 1933.

It was at this time Jung made another university appearance not mentioned anywhere in his writings or in the literature on Jung. The sole reference is in an article "Collective Guilt" that appeared in the December 27, 1945, issue of *The Nation* (Bern). In it the reporter challenged Jung's recent writings about the issue of the collective guilt of the German people for the crimes of Nazism. He says that he only heard Jung speak once, in the summer of 1933 at the University of Frankfurt am Main. Jung was introduced by the then well-respected poet Rudolph Binding who said that the excellent qualities of German youth that had previously been hidden had been realized in the victory-year 1933. Jung walked across the swastika-decorated stage and gave an address that had nothing to do with politics. He was praised by students who said, "If the swastika doesn't disturb the famous Swiss professor, then certainly everything will be in order."

Until something turns up in either Jung's unpublished papers or in the university's archives, we must rely on inferences to understand the highly suggestive lead that is presented here. First we have to consider the master-of-ceremonies, Rudolph Binding (1867–1938). Literary success had come to him only at the age of forty after the publication of his work by Grand Duke Ernst Ludwig. His service as a reserve officer in a cavalry regiment and on the Western Front led him to cultivate a chivalric code of honor that was modeled on that of the English "gentleman." He attended the 1928 Kulturbund Conference in Prague where he talked about youthful idealism and the war experience; he criticized Germany's educational system for its uselessness for "life" and for its failure to instill ideals in his generation.[6] He received an honorary doctorate from the University of Frankfurt in 1927 and the Goethe Medal in 1932. At the time of Jung's visit, Binding had become an apologist for the German revolution in his *Reply of a German to the World.* His personal connection to the University of Frankfurt and his well-known opposition to conventional academic pedagogy made Binding an attractive choice to introduce Jung.

Sweeping changes were underway at German universities as administrators and professors unacceptable to Nazi student groups and educational authorities were dismissed and replaced by those considered reliable. The most notorious case from this period was that of Heidegger's assumption of the rectorship at the University of Freiburg im Breisgau.[7] Also important was the conduct of two other Nazi-appointed rectors, Alfred Baeumler at Berlin University (and a supporter of the book-burning there) and Ernst Krieck at the University of Frankfurt.[8] Born in 1882 Krieck achieved success in 1921 with his *Philosophy of Education* and became a professor at the Pedagogical Academy in Frankfurt. Like Binding, Krieck had contributed to Keyserling's *Der Leuchter*; his "Aufgabe der deutschen Bildung" ("The Task of German Developmental Education") appeared in the 1930 issue along with a contribution by Leopold Ziegler. He joined the Nazi Party in 1932 and aspired to be Nazi Germany's educational Führer. To that end he founded and led the Cultural-Political Working Community of German University Teachers that lobbied for educational reforms along Nazi lines.

In his address "The Renewal of the University" Krieck said

> the University of recent decades has no longer had a unified basis and a singular direction of meaning. It has become a mere assemblage of dozens of specific disciplines, ruled by a false understanding of autonomy and independence. What is required in this situation is more than a practical reorganization. The universities must be given a new meaning and purpose. This is to be found only in a "unified folkish-political worldview."[9]

Committed to a renewal of German society, Krieck popularized his views in a journal whose title *Volk im Werden* (*A People in the Making*) was taken from his 1932 book by that name.

Jung's "Vom Werden der Persönlichkeit" echoed Krieck's title but differed from it in a fundamental way. Krieck put the primary value on the collective while Jung was arguing for the value of individual personality development. This position was generally out of favor with Jung's student audience. Based on his own temperament and approach to psychology, he emphasized the "inner" aspect of the individual personality and so offered his advice to those uncertain about how to respond to the mobilization going on around them. We do not know exactly what Jung said to the students that day in Frankfurt. It is likely it was based on the lecture he had been giving frequently in Germany. We should remember they would have been hearing the following observation. "There are times in the world's history—and our own time may be one of them—when good must stand aside, so that anything destined to be better first appears in evil form" (CW 17, p. 185). Jung undercut his defense of the individual in this surrender of the moral imperative.

This statement captured Jung's attitude toward political developments in Germany and is derived from two important influences on his intellectual development. The first was nineteenth-century German Pietism that stressed an "inwardness" justified by Christ's maxim "Render unto Caesar the things that are Caesar's and to God the things that are God's." This strategy of withdrawal was diametrically opposed to the activism of the Confessional Churches then forming in Germany to resist Nazi pressure on the organized churches. The second comes from the broadened sense of man's spiritual history that Jung got from his study of Taoism with Richard Wilhelm. It recalls the process of withdrawal followed by a Chinese sage during one of that country's periodic "times of trouble." It brings to mind the phrase "inner emigration" coined by Gottfried Benn to explain his decision to live and work in Nazi Germany. Ironically, Jung did not withdraw (or "stay within the mandala") but chose to become an active and vocal presence in Nazi Germany.

In November Jung was invited by the Kant Society of Bonn to give a lecture at the university. Again he lectured on "The Development of Personality" and was introduced by Erich Rothacker.[10] Rothacker held a chair of philosophy with an emphasis on psychological investigation and had developed a characterological model of the psyche that was indebted to Freud but that rejected the instinct theory.[11] It was around this time that Rothacker became the head of the *Volksbildung* department of the Propaganda Ministry. He was a member of the German Philosophical Society whose growing membership reflected a conservative orientation in both philosophical and political terms. Since 1927, the organization's president had been Felix Krueger, Wilhelm Wundt's successor at Leipzig and a leading critic of the Gestalt School. Krueger criticized Gestalt psychology for its neglect of the pre-rational feelings that affected a person's total functioning and developed an alternative theory of *Ganzheit* ("Wholeness") to account for them.[12] Krueger maintained his university position during the Nazi era, adjusting his concepts to Germany's new political realities. Among the students he attracted was Karlfried Graf von Dürckheim who had also studied with Hans Freyer, the conservative sociologist whose *Revolution From the Right* (1932) provided intellectual justification for the overthrow of the Weimar Republic.[13] An interest in Jung led Dürckheim to study Eastern spiritual practices and to a professional affiliation after World War II with a group of Jung-influenced individuals that included Gustav Richard Heyer.[14]

Jung's "On Psychology" was published in the May and June issues of the *Neue Schweizer Rundschau.* Of particular interest are the other articles that appeared with it. The lead article in May was "The Swiss and National Socialism" by Albert Oeri, Jung's old friend, editor of the *Basler Nachtrichten* and member of the Swiss Parliament. He asked if the country was "fully immune from the antidemocratic infection from Germany?" While saying that the Swiss would not become "hostile to things German" he urged his readers to dedicate themselves to the "democratic conservatism" that had brought freedom and prosperity to the country. Max Rychner, the journal's former editor and currently with the *Kölnische Zeitung,* contributed "The New Germany," which took a more sympathetic attitude to events there, citing the sense of renewal that was electrifying the country. These sentiments were shared by many conservative intellectuals, Jung among them, during the Nazi regime's early years. In 1945 Jung wrote another article for this journal and said

> Our judgment would certainly be very different had our information stopped short at 1933 or 1934. At that time, in Germany as well as in Italy, there were not a few things that appeared plausible and seemed to speak in favor of the regime... after the stagnation and decay of the post-war years, the refreshing wind that blew through the two countries was a tempting sign of hope.[15]

Thomas Mann, a conservative supporter of the Weimar Republic and a particularly perceptive critic of this view recorded the following observations in his diaries in the fall of 1933. "There is a piece [in the *Neue Zürcher Zeitung*] by Rychner of Cologne in which he speaks of the isolation of Germany and her painful preoccupation with herself. For an analysis of this aspect of Germany, always out for different things from what the world needs, see Nietzsche."[16] The following March he noted that "The *N.Z.Z.* carries one of Rychner's pro-Nazi articles, this one about the mood of confidence and hope in Germany, whose isolation is after all the common fate of all nations in the grip of a revolution. In league with the future, etc. It sounds like a paid advertisement, and Rychner may already be an agent of the Propaganda Ministry..."[17] Mann would later also express his opinions about Jung's behavior during this same period.

The June issue was devoted to "The Fronts," those Swiss authoritarian-fascist organizations that now attracted attention with the recent Nazi success in Germany. They were the subject of a report prepared for the Simon Wiesenthal Center *Survey of Nazi and Pro-Nazi Groups in Switzerland 1930–1945* by Alan Schom (June 1998). The most influential of them as far as providing leaders for the whole movement was the New Front. It had only recently merged with the National Front and was the training ground for several figures who would go on to form new groups with close ties to Nazi Germany throughout the 1930s and 1940s. The first was Hans Oehler (1888–1967) who founded and edited the *Schweizer Monatshefte für Politik und Kultur.* He helped found The People's League for the Independence of Switzerland, which opposed Swiss participation in the League of Nations. He was a popular speaker in the student circles that formed around Robert Tobler (1901–1962), a doctor of law candidate who became the Gauführer of Zurich in 1933. After being fired as editor of the *Schweizer Monatshefte* Oehler went over to the *National Hefte* from 1933 to 1945. In 1957, he was convicted of treason in a Swiss Federal court. The third important leader of this group was Rolf Henne (1901–1961) whose father was a prominent Schaffhausen doctor and whose mother was a cousin of Emma Jung's.[18]

A.H. Wyss, the author of the article on the New-National Front, discussed the contributions of these men and enunciated an ideological rhetoric typical of that expressed in the other seven selections. He lambasted the "sterile party organizations" and the "fawning servility" to "French ideas." He championed a spirit that would counter the "Jewish spirit" and "barren intellectualism," one that was "rooted in the mysterious instincts."

The placement of Jung's article in the *Neue Schweizer Rundschau* recalls that of his articles as they appeared in the *Europäische Revue*: a psycho-cultural piece juxtaposed with others of more specifically political or economic interest. This had been recognized by Karl Näf who when awarding Jung the Zurich Literary Prize noted that "Jung is considered to have brought his psychological findings out of the consulting room and into the news columns, and in this manner exercises authority in the spiritual confrontations of our time."[19] Conservative power-brokers in Germany and Switzerland had found Jung to be an available and active spokesman for their point of view.

The Radio Berlin Interview

Jung was in Berlin from June 26 to July 1, 1933, to give a dream seminar to the local Jung Society. Barbara Hannah reports that it was during this visit that Jung had an appointment, based on a misunderstanding, to meet Goebbels that was arranged through Otto Curtius.[20] On the first day of the seminar Jung was interviewed by one of his followers Adolf Weizsäcker on Radio Berlin. In the 1920s Weizsäcker had been active in the Koengener Bund, a youth group founded by Jacob Hauer that sought to promote Germany's spiritual renewal through its conferences and magazine *Die Kommende Gemeinde* ("The Coming Community"). The Bund "looked beyond the Christian revelation and hoped to find themselves in nature, in history and in their own life experience."[21] It was part of the wider movement to establish a uniquely

Germanic form of religion. After receiving a doctorate in philosophy from Marburg Weizsäcker became interested in Jung's ideas. Magnus Llunggren notes that this led to an analysis with Emil Medtner and that the two men held "similar political views."[22] What he doesn't make explicit but becomes clear from reading his book and this interview is that these views included both anti-Semitism and support for Nazism.

Before analyzing the interview, it is important to recall the momentous events that had taken place in those first months of Nazi control: February—the Reichstag fire, March—the national plebiscite endorsing Nazi control, April—the national boycott of Jewish businesses, and May—the burning of books by university students. The process of *Gleichschaltung* was in full swing as cultural and social institutions rushed to conform themselves to the dictates of Nazi policy. A top Nazi priority was control over all German media and Goebbels' Ministry of Propaganda soon censored all scripts and approved all radio commentators. Control of language was essential for the creation of a healthy new Volk and the elimination of the pernicious ideas that had plagued Germany for too long.

Goebbels explained the world-historical significance of the book-burning saying that "the age of extreme Jewish intellectualism has now ended, and the success of the German revolution has again given the German spirit the right of way..."[23] As each author's books were consigned to the flames it was accompanied by a condemnation. "Against the overvaluation of instinctual urges that destroy the soul, for the nobility of the soul! I surrender to the flames the writings of Sigmund Freud."[24] Words were now live ammunition in the Nazi campaign to win the long-smoldering Weimar culture wars.

The theme of the interview was identified by the announcer in an opening statement not included in the English language version of the text (*CGJS*, pp. 50–66) as the contrast between Freud's "corrosive psychoanalysis" and Jung's "constructive teaching." These phrases cued listeners to the dichotomy that would structure the interview that followed. What is important to appreciate is the degree to which Jung echoed Nazi terminology in his remarks. He referred to the creation of a *Volksgemeinschaft* (national community), a term being promoted by the Nazis to characterize their goal of societal renewal. Later, when the discussion returned to his differences with Freud, Jung summarized his disapproval by characterizing psychoanalysis as a technique that was *Lebensfeindlichkeit* ("hostile-to-life"), a label used to describe anything that the Nazis felt threatened that community.

Besides "Freud" and "nationalism," Weizsäcker solicited Jung's opinions about "youth," "individualism," and "leadership." Jung elaborated on these topics using a set of stock phrases that had an almost ideogrammatic quality for him and his audience. For example, "corrosive" was linked to a cluster of other words under the rubric "intellectual" (at one point, Jung referred to the "false intellectualism" of the nineteenth century that had substituted an abstract conception of man for a realistic one). This was used to describe the Freudian and Adlerian approaches to the psyche that were not just reductive, but were "destructive' and "tearing to pieces." Weizsäcker referred to their Jewishness indirectly when he said "Dr. Jung comes from a Protestant parsonage in Basel. This is important. It puts his whole approach to man on a different footing than [theirs]." The point was not to explore differences in the theories, but to underscore Jung's Christian heritage and to demonstrate his sensitivity to Germany's spiritual history.

Jung described German nationalism as a "nation-building force" in comparison to the chauvinism of the Western European countries. He attributed this difference to "the youthfulness of the German nation." This concept of Germany as a youthful nation had been developed by Moeller von der Bruck, one of the Weimar Republic's leading neoconservative critics and the popularizer of the term *Third Reich*. The following comment is far more troubling when we remember that it was made just six weeks after some of these "youthful Germans" had burnt books by Sigmund Freud.

> The assurance of German youths in pursuit of their goal seems something quite natural to me. In times of tremendous movement and change it is only to be expected that youth will seize the helm, because they have the daring and drive and sense of adventure. After all, it is their future.

Toward the end of the interview Weizsäcker asked Jung his opinions about the idea of personal leadership and a leadership elite as it was developing in Germany.

> Times of mass movement are always times of leadership. Every movement culminates organically in a leader, who embodies in his whole being the meaning and purpose of the popular movement. He is an incarnation of the nation's psyche and is its mouthpiece. He is the spearhead of the phalanx of the whole people in motion. The need of the whole always calls forth a leader regardless of the form a state may take. Only in times of aimless quiescence does the aimless conversation of parliamentary deliberations drone on, which always demonstrates the absence of a stirring in the depths...It is perfectly natural that a leader should stand at the head of an elite, which in earlier centuries was formed by the [deleted: "feudal"] nobility. The nobility believe by the law of the nature in the blood and exclusiveness of the race.

Jung conceded much to his interviewer by accepting the Nazi rationale for its new government and so casually dismissing parliamentary democracy. The last sentence shows just how far Jung went to accommodate himself to Nazi rhetoric.

How can Jung's opinions in this interview be explained? What he said involved more than the psychotherapeutic technique of adopting a client's idiom to foster dialogue. Jung had held some of the views he expressed (like those about Freud) for many years while others were influenced by the immediate circumstances surrounding the interview. An anecdote related by Michael Fordham, later one of the senior analytical psychologists in Great Britain, can shed some light on this. In 1933 he had gone to Zurich to meet Jung. He had traveled by train through Germany in a railway car with a young Jew who was making plans to emigrate. When Fordham told Jung about this encounter, the word "Jew" acted "like a stimulus word in an association test that had hit a complex, and for about three-quarters of an hour Jung delivered a long discourse on the Jews, their history and their differences from Christians and Europeans." Because of his positive transference "It would never have occurred to me that Jung was possessed and became unrelated to the person he talked to... Subsequently, when Jung talked compulsively I concluded he was not well..."[25] This anecdote provides a clue to understanding Jung's state of mind at the time. Fordham interpreted what Jung was saying in terms of the word association test that Jung had developed. His studies had concluded that disturbances in associations were caused by complexes with an unconscious emotional core. The word Jew had triggered in Jung a series of associated thoughts whose emotional dimension was largely unconscious.

Weizsäcker addressed Jung as a Swiss observer relatively detached from events in Germany but in this and subsequent situations Jung made it clear that he was affected by what was going on there. In 1935 he told an audience in London

> I saw it coming, and I can understand it because I know the power of the collective unconscious. But on the surface it looks simply incredible. Even my personal friends are under that fascination, and when I am in Germany, I believe it myself, I understand it all, I know it has to be as it is. One cannot resist it. It gets you below the belt and not in your mind, your brain just counts for nothing, your sympathetic system is gripped.[26]

Here Jung clearly admits to a *participation mystique* with developments in Germany, a situation he would analyze in his 1936 article "Wotan" where he would interpret what was going on as a fulfillment of his 1923 prediction to Oscar Schmitz: a collective regression to Germany's pre-Christian, barbarian roots. His fascination with the deeper meaning of this phenomenon led him to underestimate the rage that propelled the Nazi movement. An example of Jung's emotional connection with Germany occurred at the time of the Röhm Purge in June 1934. He was struggling with a deep sense of oppression and staying alone at his retreat tower at Bollingen. After several days he got in touch with a colleague who told him of the purge (*CGJS*, pp. 182–183). Other factors such as flattery, opportunism, and prestige played roles in this interview. Weizsäcker addressed Jung as "the most progressive psychologist of modern times" (p. 60).

Although not mentioned during the interview, Jung had only five days earlier accepted the presidency of the International General Medical Society for Psychotherapy. Now, twenty years after his resignation from the International Psychoanalytic Association Jung was once again the leader of a psychotherapy organization. It would mean greater involvement with German affairs and an opportunity to promote his school of psychology at the expense of psychoanalysis.

Conservative Intellectuals Fall in Line

Jung was drawn early on into Nazi cultural politics, especially in the context of its attack on psychoanalysis. His reputation made him the obvious candidate for critics of psychoanalysis to back in their quest for a non-Freudian psychotherapy. In the May 14, 1933, issue of the *Berliner Börsen-Zeitung* a group of essays appeared under the collective title "Against Psychoanalysis": they included "Christianity and Psychoanalysis" by Frank Mauran; "The Conquest of Psychoanalysis" by Hans Kern; "Psychoanalysis and True Soulfulness" by Felix Krueger; and "The Reform of Psychoanalysis through C.G. Jung" by Christian Jenssen. None of them explicitly mention Freud's Jewish identity but criticize him from different angles: Mauran emphasized the incompatibility of psychoanalysis and Christianity; Krueger claimed that it lacked scientific credibility. Kern had been active for ten years in cultural journalism mostly through his involvement with Ludwig Klages; his articles appeared frequently in the *Zeitschrift für Menschenkunde* among other publications and his book on Carus was published by the Niels Kampfmann Verlag in 1926. His article was devoted to establishing a German pedigree for the concept of the unconscious, one that started with Goethe and Herder, was developed by the Romantics, and continued through Nietzsche to Klages. Psychoanalysis was dismissed as a relic of nineteenth-century materialism, an irreligious "mechanism of the unconscious." A notice announced that another article "Psychoanalysis and Marxism" was planned.

Christian Jenssen was the Nordic pseudonym of one Gottfried Martin. Born in 1905, he was an editor and art critic in Cologne, possibly with the *Kölnische Zeitung*. His career demonstrated the opportunities created by the policy of *Gleichschaltung*. He became the literary agent of Hans Blunck, a fellow North German who became one of Nazi Germany's most celebrated government-sponsored writers. Blunck's work expressed a deep attachment to the land and people of his native region and included many poems and stories about Germanic prehistory. His prolific output was regularly published in a variety of editions, many of them for use in German schools. Jenssen edited the four-volume edition of Blunck's Collected Works that came out in 1941 and then continued a literary career into the 1950s.[27]

Jenssen's article contrasted Freud and Jung using a formula acceptable to Nazi censors.[28] The "blood-conditioned mentality of Freud and Adler" created one-sided theories that were "sharply-drilling intellectualism." "Freud's creed" was "an alien element in the German nation." He supported this critique with a quote (later deleted) from Jung's "On the Psychology of the Unconscious." "The sexual theory is unaesthetic and intellectually less satisfying, the power theory (Adler) is decidedly poisonous."[29] Jenssen then made his case for Jung whose name has been "studiously withheld" from the public. Jung was a "true Swiss-German, conservative by nature . . . a German thinker." Jung's approach, the product of a decidedly German mentality, overcomes doctrinaire intellectualism by helping the patient discover the creative powers of the unconscious. Jenssen went on to outline the main features of Jung's theory: archetypes (persona, anima/animus) and collective unconscious, introversion/extraversion, individuation, and the Self.

The article employed the bluntest contrasts to make its point: Freud—alien, destructive, fragmented; Jung—native ("genuine Swiss German, conservative by nature"), constructive, holistic. He also relied on innuendo when he refers in passing to Jung's name having been studiously withheld from the public. The implication was that this was the result of a deliberate policy of a Jewish press sympathetic to the Freudian position. This figured in Jung's letter of thanks to Jenssen on May 29.[30] Jung wrote that he appreciated the general tone of the article since his work was unknown in Germany and few people knew that he was saying something different from Freud.

> My scientific conscience did not allow me, on the one hand, to let what is good in Freud go by the board and, on the other, to countenance the absurd position [more accurately, "distortion"] which the human psyche occupies in his theory. I suspected at once that this partly diabolical sexual theory would turn people's heads and I have sacrificed my scientific career in doing all I can to combat this absolute devaluation of the psyche.

Here in a letter written less than three weeks after the Nazi book-burning Jung used inflammatory language that more than matched that of his correspondent. Although expressed privately, it expressed in embryo the critique of psychoanalysis Jung would soon publish in "The State of Psychotherapy Today." Although always acknowledging its legitimate but limited applicability,

Jung characterized psychoanalysis in highly charged, emotional language. His use of "sacrifice" and "combat" owed something to the rhetoric emanating from Germany. Jung would remain oblivious to the fact that the basic ground rules in his debate with Freud had changed and that what he had to say about Freud would now find a highly partisan audience in Germany.

Jung felt no need to qualify Jenssen's characterization and had only one correction to make, declaring that he did not consider himself to be from the Freudian school. He was caught up in the flattery that assuaged his hurt feelings regarding lack of recognition. It also appealed to his ambition to promote his school of psychology. Jung would be aligning himself with forces that had the legal if not the intellectual power to put psychoanalysis in its place. Speaking in his own defense after the war, Jung spoke about how his warning voice about developments in Nazi Germany had gone unheeded at the time. While it is true that Jung did raise a warning voice in the early 1930s, this letter reminds us that his warnings pertained to the dangers of psychoanalysis and not to the dangers of Nazism.

As Jewish and leftist writers were arrested or driven out of the country, a group of conservative intellectuals came forward to explain the German Revolution to the outside world. They believed that the Nazis were a force that could be contained by the traditional power elites and that any excesses they perpetrated were temporary aberrations necessary to secure "law and order."

One example of this genre was *Sechs Bekenntnisse zum neuen Deutschland* (*Six Testimonials to the New Germany*), a booklet published by the Hanseatische Verlag. It opens with a letter of Romain Rolland to the editor of the *Kölnische Zeitung* and was printed on May 21; Rolland challenged the editor to recognize the fact that the "national-fascist Germany is the worst enemy of the true Germany," and of freedom and creativity. Writing days after the book-burning, he expressed concerns about the "autodafes [*sic*] of ideas" and the "outrageous interference in the politics of the universities." Among the respondents was E.G. Kolbenheyer, a longtime Nazi supporter and author of a best-selling trilogy about Paracelsus, the great Renaissance physician who was being promoted as an exemplar of German science. Another was Wilhelm von Schloz who, like Kolbenheyer, was active in the cultural politics of the "coordinated" Prussian Academy of Literature.

We have already met another contributor, Rudolph Binding. His response was phrased in the patriotic terms he had been using since the war. Rolland, he said, could not understand the effect that the "dictated" Treaty of Versailles had had on the German people, "The world hasn't lived what we have lived." The yearnings of the German people were "natural, not political" and "inner, not outer." He dismissed the importance of the forced emigrations and corrected Rolland's idealization of German culture, saying that the essence of German genius lay in its universality, not in its internationalism.

The theme of Germany's spiritual renewal was also evident in the contemporary writings of Gottfried Benn (1886–1956), one of the country's leading Expressionist poets, and a man who shared Jung's avant-garde conservative sensibilities. His work, strongly influenced by Nietzsche and his experiences as a doctor, conveyed the brutality and meaninglessness of modern existence. Like other intellectuals he dismissed the Weimar Republic for its banality and lack of authenticity. In 1932 he was elected to the Prussian Academy of Literature to which he delivered an address, "After Nihilism." He sent it to Max Rychner who wrote about it in the November 29 issue of the *Kölnische Zeitung*.[31]

Benn's activities as leader of the Prussian Academy of Literature Poetry Section during the process of *Gleichshaltung* prompted a letter from Klaus Mann, the son of Thomas Mann, who had resigned from the academy in protest and was in exile.

> I now learn for a fact that you—as indeed the only German author our kind had counted on—have not resigned from the Academy... What company do you keep there? What could induce you to put your name—to us a byword for high standards and an all but fanatic purity—at the disposal of men whose lack of standards is unmatched in European history and from whose moral squalor the world recoils?... It seems almost a law of nature that strong irrational sympathies lead to political reaction if you don't watch out like the devil. First comes the grand gesture against "civilization"—a gesture I know as only too attractive to intellectuals, then, suddenly, you've reached the cult of force, and the next step is Adolf Hitler... I know a man need be no obtuse 'materialist' to want what is reasonable, and to loath hysterical brutality with all his heart.[32]

Benn wrote an open letter in reply that was given widespread coverage by Joseph Goebbels. It soon appeared as "Answer to the Literary Emigrants" in a booklet *Die neue Staat und die Intellektuellen*. He tried to convey his feelings about what he considered a unique historical moment. Considering the charge of "barbarism" he said that

> You put it as if what happens in Germany now was threatening culture, threatening civilization, as if hordes of savages were menacing the ideals of mankind as such. But let me ask you in turn: how do you visualize the movement of history? Do you regard it as particularly active at French bathing beaches?... you would go farther if you discarded this novelistic notion and viewed it more as elemental and impulsive, an inescapable phenomenon... this view of history is not enlightened and not humanistic, it is metaphysical, and even more so is my view of man... one that considers him as mythical and deep... for man is older than the French Revolution and more stratified than the Enlightenment believed... he is eternal Quaternary, a horde magic feuilletonisticallly decking even the late Ice Age, a fabric of diluvial moods, Tertiary bric-a-brac; in fact, he is eternally primal vision.[33]

Benn went on to make his case using racial imagery involving the preservation of the white race threatened by Negro colonial troops, which conjured up memories of their role in the French occupation of the Rhineland after World War I. He spiced it up with personal insults ("your bourgeois nineteenth century brain") and sarcasm by suggesting that those driven from Germany spent their time lounging at seaside resorts.

The extent to which some of this sounds like Jung is striking. Besides a generally conservative view of European culture (Benn quoted Burckhardt, Fichte, and Nietzsche), Benn and Jung were both doctors who felt their professional experiences gave them an advantage over their intellectual opponents whose bookishness insulated them from life's gritty realities. Recall Jung's 1923 letter to Oscar Schmitz in which he talked about cultivating the barbarian component in the German psyche.

> I am a doctor, and am therefore condemned by my speculations under the juggernaut [more accurately, the "Wheel"] of reality, though this has the advantage of ensuring that everything lacking in solidity will be crushed. Hence I find myself obliged to take the opposite road from the one you appear to be following in Darmstadt [that is, at the School of Wisdom].[34]

Both men diagnosed their German "patient" as regressing to a deep, archetypal/primordial state that could be the source of healing for the collective neurosis that originated in the process of secularization. They were slow to see the psychotic rage that was driving events there.

As the Nazis consolidated their control over Germany, publishers deemed acceptable to the authorities quickly adapted to the new situation. As their Jewish competitors were eliminated they found their new opportunities circumscribed by the need to conform to the guidelines of Nazi censorship. Goebbels was, however, careful to permit the continued publication of a wide range of nonideological literature to maintain the illusion of normalcy in German cultural life. Publications aimed at the intelligentsia stressed the continuity of recent developments with their antecedents from the Romantic School who were celebrated for challenging the worldview of the Enlightenment and French Revolution. Honored as pioneers in the discovery of the unconscious, their work was held up as an inspiration for a new generation of German artists and scientists.

Typical of this kind of writing was Otto Kankeleit's *Die schöpferische Macht des Unbewussten: Ihre Auswirkung in der Kunst und in der moderne Psychotherapie* (*The Creative Power of the Unconscious: Its Effects in Art and in Modern Psychotherapy*) (Berlin: Walter de Gruyter, 1933). "We are only a breadth in God's mouth" is the motto of his introduction that contrasts the one-sided intellectual position of civilization with the possibilities of true culture. He then cited Edgar Dacque and Ludwig Klages to support his argument about the role of symbols in creativity. The book consisted of the responses of eighteen individuals to a set of six questions that dealt with the influence of the unconscious on creativity, including the role of dreams and stimulants. The lead position went to Hans Blunck, the Nazi literary lion who was about to be made president of the Reich Chamber of Authors, a position that included reviewing lists of blacklisted authors. About the same time, he founded the Speech-fostering Office to combat the infiltration of foreign words by promoting the use of old German words.[35] Rauschning remembered hearing Blunck's "flowery oration" at the opening of a cultural event in Danzig.

> Statements were solemnly made that the original culture of mankind had not arisen around the Mediterranean at all, but on the shore of the Baltic, created by the Nordic races. The Baltic was the home of heroism and Aryan racial culture, and the Mediterranean was the seat of racial decay and Semitic degeneration... Hitler is not interested in the pure Aryan blood of the Scandinavians, nor in the northern myths of Viking heroism. He is interested in the iron-ore mines... Herr Blunck, and our Swedish friends are playing gratuitous parts in a play the background of which they have never seen.[36]

Among the other contributors to the anthology were Count Keyserling, Alfred Kubin, and Carl Jung who would have known Kankelheit through their membership in the General Medical Society for Psychotherapy. This fact is not identified in the General Bibliography of Jung's works (CW 19), which lists only the 1959 second edition (the text of which appears in CW 18, paragraphs 1760–1768). Both editions are listed in the catalogue of Jung's personal library and a comparison of the two reveals interesting continuities and revisions. To his own reprinted foreword Kankeleit added a foreword by Jung and fifty-six additional contributors who represented the whole gamut of postwar intellectual life. Among them were Viktor Frankl, death-camp survivor and founder of logotherapy, and the American writer and humanitarian Pearl S. Buck. There were two prominent Jewish Jungians Jolande Jacobi and Erich Neumann who had likely been recruited by Jung himself. Another group was composed of those who had held prominent positions in Nazi literary and scientific life. This included Eugen Fischer and Fritz Lenz who with Erwin Bauer coauthored *The Principles of Human Heredity and Race-Hygiene* (1921), which had a decisive influence on Hitler's outlook.[37] They later helped implement these ideas in Nazi legislation and research programs whose goal was the elimination of those deemed undesirable by the state. These leaders in anthropology and genetics provided the scientific rationale for an ideology that demanded the sacrifice of the inferior for the greater good of the racial community. Its first targets were the mentally ill who had been targeted by race-hygienists as an unnecessary economic burden on society. In 1920, Karl Binding, a jurist and father of Rudolph, collaborated with psychiatrist Alfred Hoche on *The Sanctioning of the Sacrifice of Lives Unworthy to be Lived*. The first practical step in the euthanasia program was legislation authorizing the sterilization of the mentally ill.[38] This leads us to the most significant deletion from the original text. Kankeleit's original introduction concluded by saying that his subject matter invited reflection on the relationship between psychotherapy and racial hygiene and recommended his own early works, which included such works as *Sterilization on Racial-Hygienic and Social Grounds* (1929) from Lehmanns Verlag.

The International General Medical Society for Psychotherapy

On April 6, Ernst Kretschmer resigned as the General Society for Medical Psychotherapy's president. Its membership was predominately German, the majority of whom shared a conservative-nationalist philosophy. Its main purpose was to promote psychotherapy as a medical specialty distinct from psychiatry and neurology. Its congresses and journal *Der Zentralblatt der Psychotherapie und Ihre Grenzgebiete* sought to achieve this goal through a policy of theoretical eclecticism and legislative lobbying. This brief synopsis is based on Cocks' *Psychotherapy in the Third Reich*, which contains the most comprehensive history of the Society.[39] My account will focus on Jung's personal relationship with the conservative-nationalist group that came to dominate the organization, especially that with one of its leaders, Jung's "crown prince" Gustav Richard Heyer. It will also consider Jung's professional conduct as president of the reorganized International General Medical Society for Psychotherapy.

When Jung joined the organization in 1928 he got reacquainted with some of his old colleagues from Munich who had been active in the formative years of the psychoanalytic movement. "Leonard Seif (1866–1949) founded a Freudian group in Munich 1911, separated from psychoanalysis in 1913, met Alfred Adler in 1920 and thereafter became a leading figure in the Society of Individual Psychology. Adler broke with him in the 1930's after Seif's group compromised with the Nazis."[40] Another alumnus of the Munich chapter of the early psychoanalytic movement was Hans von Hattingberg, who along with Arthur Kronfeld and J.H. Schultz had been identified as a "wild analyst" by Karl Abraham.[41] A psychiatrist, he was at the time publishing a series of books on the neuroses and psychoanalysis with Lehmanns Verlag, which also published the works of Gustav Richard Heyer. Their works fit nicely into the booklist of a firm that was one of Germany's leading publishers of medical texts.

By the 1930s Lehmanns Verlag had also become Germany's leading publisher of racial hygiene literature. This policy reflected the convictions of Julius Lehmann, the firm's founder, an early and vocal supporter of the Nazi Party.[42] In 1940 the firm issued a fiftieth anniversary volume that recounted its history and included a bibliography of such leading racial hygienists as Hans Günther, Albert Hoche, and Ernst Rüdin. The book included their photos as well as those of von Hattingberg and Heyer (the same one that had been used in the 1935 Eranos Yearbook). Another photo was that of Erwin Liek, a leading figure of the German natural health movement and founder of its journal *Hippokrates*. One of its contributors was Heyer who had been introduced to it by his wife Lucy Grote. She was a physical therapist and daughter of Dr. Louis R. Grote who became a chief of Nazi Germany's leading natural health clinic in Dresden, which was named after its patron Rudolph Hess. What is interesting about Liek's photo is the fact that the rune of death rather than a cross was used to symbolize the fact that he was deceased. This attests to the degree to which natural medicine had become entwined with neo-paganism. This relationship had its roots in the turn-of-the-century Occult Revival. German occultism was heavily racialized by such groups as the Thule Society founded in Munich in 1917, which included among its members the publishers Eugen Diederichs and Julius Lehmann as well as Dietrich Eckart, Adolf Hitler's most important mentor.[43]

Cocks tracks the political maneuvering of the General Medical Society's leadership vis-à-vis the psychiatric establishment (with its bias against psychotherapy) and the nascent Nazi health bureaucracy. With Kretschmer's resignation, Jung as the Society's honorary vice president agreed to assume the presidency, an offer he accepted with the understanding that he would be the president of a reorganized International Society to be composed of different national groups. This was a bow to the realities of the policy of *Gleichschaltung* in Germany where the Society's German members were in the process of organizing themselves into the German Medical Society for Psychotherapy under the leadership of Matthias Göring, a cousin of Herman Göring. Their executive committee included Cimbal, Häberlin, von Hattingberg, Heyer, Künkel, Schultz, Schultz-Hencke, Seif, and Viktor von Weizsäcker.[44]

Jewish members such as Walter Eliasberg and Arthur Kronfeld were eliminated from leadership positions and their departures given only passing reference in the *Zentralblatt*. ("The former editors have resigned from the staff of the Journal.") This is yet another example of the self-censorship that had come to prevail in Nazi Germany.[45] Some of these expelled members were resourceful enough to start an alternative journal *Psychotherpeutische Praxis* as a forum for medical therapists barred from other publications on racial or political grounds. Jung discussed this situation in a letter to Rudolph Allers of Vienna.

> A foreign editor, I fear, would in the present circumstances meet with not a few difficulties, because the German government, as you know, seems to like having the editors of all periodicals appearing in Germany in safe and uncomfortable proximity. Otherwise I would have proposed you as editor. I have already written to Cimbal on this matter but as yet have received no answer. It must unquestionably be a "conformed" editor, as he would be in a far better position than I to have the right nose for what one can say and what not. In any event it will be an egg-balancing dance. Thank you very much for sending me the announcement of this new journal [footnote: "Probably *Psychotherapeutische Praxis*"]. I have declined with thanks to cooperate because I propose to turn my interest more to the *Zentralblatt*. Psychotherapy must see to it that it maintains its position inside the German Reich and does not settle outside it, regardless of how difficult its living situation there may be.[46]

The strident anti-Semitic rhetoric of the Nazis was accompanied at first by a gradualist policy toward Jews that aimed at their segregation from the German Volk. Legislation in April deprived thousands of civil servants (including teachers and professors) of their jobs and large Jewish-owned businesses were "Aryanized." Jews lived with the constant threat of intimidation and detention; thousands emigrated and those that remained were forced to adjust to the tightening legislative constraints. Contributions by Jewish writers to most publications ceased and their names disappeared from publications with which they had been affiliated.

One such individual was the Viennese gynecologist Bernard Aschner (1883–1960), an advocate of a holistic approach to human health, who served on the editorial board of *Hippokrates*. Jung became acquainted with Aschner's work in the 1920s and was struck by its similarities to his own.

> Purely intellectualistic, analytical, atomistic, and mechanistic thinking has, in my opinion, landed us in a *cul de sac*, since analysis requires synthesis and intuition. The humoral pathology of Aschner, who, incidentally, has rediscovered medical techniques based predominately on intuition through his translation of Paracelsus, is for me proof that the most important insights into body and mind can be gained by ways that are not purely rationalistic.[47]

Aschner, a Jew, was dedicated to a medical-scientific approach that the Nazis were in the process of claiming was an exclusively "German" domain. He suffered ostracism and after 1938, exile.[48] Theodore Lessing (1872–1933), another Jewish intellectual, suffered a more tragic fate. As a schoolboy he had developed a friendship with Ludwig Klages with whom he shared a fondness for *Lebensphilosophie*; Lessing expressed this in his book *The Decline of the*

Earth by the Spirit (1924) and in his editorial work on a new edition of Carus' *The Symbolics of the Human Form* for Niels Kampfmann (1925). He had never been forgiven for his criticism of Hindenburg's conduct during World War I and left Germany for Prague soon after Hitler's assumption of power. Unfortunately, he did not go far enough since Nazi agents soon tracked him down and assassinated him.[49]

Gleichschaltung had created an unprecedented period of change in Germany. Jung's presidency was announced on June 21, a date apparently chosen because it was the summer solstice and therefore an auspicious day (the fact that it fell on Wednesday, Wotan's Day, would have made it seem an even more appropriate choice). Jung became active in drafting bylaws for the Society's new international structure that would be ratified the next spring at the Bad Nauheim conference. This process involved the formation of national groups in Sweden, Denmark, Holland, Switzerland, and later, Great Britain. It was necessitated by the founding of the German Society on September 15. To the Swede Poul Bjerre (1876–1964), another of his old psychoanalytic colleagues, Jung wrote,

> If we succeed in organizing some national groups in neutral countries, this will act as a counter-weight and at the same time afford the Germans a much needed opportunity to maintain a connection with the outside world in their present spiritual isolation. This connection is essential for the continued development of psychotherapy in Germany, since at present she is even more cut off than during the war.[50]

The first group to form, and numerically the largest after the Germans, was the Dutch under the leadership of J.H. van der Hoop.

Jung's letter reminds us that efforts to form groups initially took place only in those countries that had been neutral in World War I. Doctors less prone to old animosities had become active in the Society when it had first been formed. Reference to a "national group" should not obscure the fact of just how few individuals were actually involved in this effort. Cocks notes that at the 1934 Bad Nauheim conference there were seventy-one German participants, two from Holland, and one each from Sweden and Switzerland.[51]

The Zentralblatt Controversy

Due to the unsettled political situation the Society's conference planned for Vienna in April had been cancelled and publication of the *Zentralblatt* had been delayed. In December, however, its third and final number of 1933 was published. Jung in his new role as president and editor wrote a foreword that triggered a controversy that has clouded his reputation ever since. The most widely quoted passage was the following "The differences which actually do exist between Germanic and Jewish psychology and which have long been known to every intelligent person are no longer to be glossed over... I should like to state expressly that this implies no depreciation of Semitic psychology."[52]

To his critics this statement is prima facie evidence of his anti-Semitic collaboration with the Nazi government. In his defense, sympathizers emphasize the quote's explicit rejection of a negative valuation of Jewish psychology and his ongoing efforts to reorganize the Society to accommodate Jewish members threatened with expulsion from the German group. His defenders invoke his long-held position (mentioned in his foreword) that a recognition of the therapist's "personal equation" was of critical importance in the success of the therapeutic relationship. Over the years most of them have come to concede the poor timing of his remarks and his political naiveté but have not bothered to clarify what Jung actually intended by his distinction between Germanic and Jewish psychology. This is due, in part, to the linguistic ambiguity involved in the use of the word "psychology." It can either be used in a general sense to mean "mentality' or more specifically to mean "mental science." Although the first is the one that his defenders have used in defending his position it is clear from the rest of the foreword, from what he had previously written and would soon write, that Jung was referring to a Jewish mental science known to the world as psychoanalysis.

The controversy was initiated by a Swiss psychiatrist Dr. Gustav Bally in an article "Deutschstämmige Psychotherapie?" ("German-racial Psychotherapy?"), which appeared in the *Neue Zürcher Zeitung* on February 27, 1934. He wrote

> How does he [Jung] want to tell the difference between Germanic and Jewish psychology anyway?... Jung doesn't reveal to us by which method we should carry out this distinction, nor which specific value we may expect from the consideration of the racial in psychology... One who introduces the racial question in his capacity as the editor of a "gleichgeschaltet" journal must know that his words raise themselves against the backdrop of organized passions which already provide the meaning contained implicitly in his words.

The newspaper published Jung's rejoinder on March 13 and 14 followed by a postscript on the 15. Jung went to great lengths to defend himself against the accusation that he was the editor of a *gleichgeschaltet* professional journal. This misunderstanding was caused by the appearance of a statement by Matthias Göring that immediately followed Jung's foreword. It was referring explicitly to the German not the International Society when it said that

> It is particularly concerned with those physicians who are willing to promote and to practice a psycho-medical therapeutic in terms of the Nationalist Socialist world view [and that] its members who are either engaged as writers or speakers

> have read through conscientiously and thoroughly Adolf Hitler's basic book *Mein Kampf* and that they accept it as fundamental. They wish to participate in the work of the people's chancellor, in order to educate the German people for their heroic and self-sacrificing role.

This was followed by a report by Walter Cimbal that discussed the many developments that had taken place in the Society and *Zentralblatt* since the national revolution had begun. Jung pointed out rightfully that Bally had confused the International with its German affiliate but admitted that the publication of Göring's statement had caught him by surprise since it had been his understanding that it would appear in a special supplement for distribution only within Germany.

In his response to Bally Jung not only clarified a misunderstanding but made a deeply personal statement about his feelings and intellectual position. One of his objectives was to explain his motives for becoming the Society's president:

> should I—as I was well aware—risk my skin and expose myself to the inevitable misunderstandings which no one escapes who, from higher necessity, has to make a pact with the existing powers in Germany?...I have seen too much of the distress of the German middle class, learned too much about the boundless misery that often marks the life of a German doctor today, know too much about the general spiritual wretchedness to be able to evade my plain human duty under the shabby cloak of political subterfuge.[53]

Jung was quite clear about the course of action that should be followed in reorganizing the Society. "They [doctors] must learn to adapt themselves. To protest is ridiculous—how to protest against an avalanche? It is better to look out. Science has no interest in calling down avalanches; it must preserve its intellectual heritage even under changed conditions." He repeats this a little later on. "We are neither obliged nor called upon to make protests from a sudden access of untimely political zeal and thus gravely endanger our medical activity. My support for the German doctors has nothing to do with any political attitude."[54] Saying that "Martyrdom is a singular calling for which one must have a special gift," Jung deliberately chose a pragmatic, accommodationist strategy regarding *Gleichschaltung.*

What comes across here is Jung's conviction that what was happening was a natural force so powerful ("an avalanche") that individual resistance was futile. To support his view about the inevitability of current events Jung made an historical analogy that sounded like an apology for what was going on in Germany.

> We in Switzerland can hardly understand such a thing, but we are immediately in the picture if we transport ourselves back three or four centuries to a time

> when the Church had totalitarian presumptions. Barbed wire had not been invented then, so there were probably no concentration camps; instead, the Church had large quantities of faggots...As the authority of the Church fades the State becomes the Church, since the totalitarian claim is bound to come out somewhere. Russian Communism has therefore, quite logically, become the totalitarian Church...No wonder National Socialism makes the same claims! It is only consistent with the logic of history...[55]

Jung's analogizing took a sarcastic turn when he began to discuss the trouble Galileo had with the Church authorities. "Galileo had the childlike eyes of the great discoverer and was not at all wise to his *gleichgeschaltet* age. Were he alive today he could sun himself on the beach at Los Angeles in company with [Mister] Einstein and would be a made man, since a liberal age worships God in the form of science."[56] This is reminiscent of Gottfried Benn's snide comment about German émigrés lounging around seaside resorts made a year earlier.

In the last part of the article, Jung elaborates on his interest in the "imponderable differences" that exist between different groups. ("Why this ridiculous touchiness when anybody dares to say anything about the psychological differences between Jews and Christians? Every child knows that differences exist.") His sarcasm continued to be evident in the following remarks:

> I am also quite ready to suppose that I am a bigoted Swiss in every respect. I am perfectly content to let my psychological confession, my so-called "theories," be criticized as a product of Swiss wooden-headedness or queer-headedness, as betraying the sinister influence of my theological and medical forbears [*sic*], and, in general, of our Christian and German heritage, as exemplified for instance by Schiller and Meister Eckhart. I am not affronted when people call me "Teutonically confused," "mystical," "moralistic," etc. I am proud of my subjective premises, I love the Swiss earth in them.[57]

This created an opening for what followed.

> May it not therefore be said that there is a Jewish psychology too, which admits the prejudice of its blood and its history? And may it not be asked wherein lie the peculiar differences between an essentially Jewish and an essentially Christian outlook? Can it really be maintained that I alone among psychologists have a special organ of knowledge with a subjective bias, whereas the Jew is apparently insulted to the core if one assumes him to be a Jew?...I must confess my total inability to understand why it should be a crime to speak of "Jewish" psychology.

Up until this point Jung had spoken of Jewish psychology in its general sense of mentality but then switched to its specific meaning as mental science.

"I attack every leveling psychology when it raises a claim to universal validity, as for instance the Freudian and the Adlerian. All leveling produces hatred and venom . . . " This dubious assertion would be amplified in the first article Jung would contribute to the *Zentralblatt* after his becoming president.

Thomas Mann, who was now living in exile in Switzerland, noted in his diary on March 14: "C.G. Jung's self-justifying article in the *Neue Zürcher Zeitung* is most unpleasant and disingenuous, even badly written and witless; strikes the wrong pose. He ought to declare his 'affiliation' openly." A year later Mann wrote:

> If a highly intelligent man like Jung takes the wrong stand, there will naturally be traces of truth in his position that will strike a sympathetic note even in his opponents . . . His scorn for "soulless rationalism" has a negative effect only because it implies a total rejection of rationalism, when the moment has long since come for us to fight for rationality with every ounce of strength we have. Jung's thought and his utterances tend to glorify Nazism and its "neurosis." He is an example of the irresistible tendency of people's thinking to bend itself to the times—a high-class example . . . He swims with the current. He is intelligent, but not admirable. Anyone nowadays who wallows in the "soul" is backward, both intellectually and morally. The time is past when one might justifiably take issue with reason and the mind.[58]

Jung closed his reply to Bally with the rebuttal that "I did not speak of it [the Jewish problem] only since the revolution; I have been officially campaigning for criticism of subjective psychological premises as a necessary reform in psychology ever since 1913."[59] To support this, Jung cited in his afterword what he had written in his 1918 article "The Role of the Unconscious" and in the footnote to *The Relationship Between the Ego and the Unconscious*.[60] For a long time, Jung had felt that what he had to say was routinely ignored but now was to experience an unexpected and unpleasant degree of scrutiny in the press and from many correspondents. One is left with the impression that Jung never really understood what all the fuss was about. He seemed oblivious to the dramatically changed circumstances that formed the context in which his old opinions took on new meanings.

The *Israelitisches Wochenblatt für die Schweiz* cited the Bally piece in a column on March 9 and published an article "Ist C.G. Jung 'gleichgeschaltet'?" by Dr. B. Cohen exactly one week later. It was a sympathetic but not uncritical defense of Jung: "in Berlin of all places the Jungian school makes a great effort to objectively understand Biblical-Jewish thought . . . " and

> It seems quite inappropriate, in view of single mistakes due to his being inadequately informed due to mistakes caused by his application of specifically Christian standards, to discount the greatness and significance of such a

scientist as C.G. Jung due to his political orientation and even to impute base motives for his political orientation.

Ten days later, Jung wrote Cohen a letter in which he said

I am absolutely not an opponent of the Jews even though I am an opponent of Freud's. I criticize him because of his materialistic and intellectualistic and—last but not least—irreligious attitude and not because he is a Jew. In so far as his theory is based in certain respects on Jewish premises, it is not valid for non-Jews.[61]

Jung had said in his foreword that that he intended no depreciation of Semitic psychology and wrote in his rejoinder that "It seems to be generally assumed that in tabling the discussion of ethnological differences my sole purpose was to blurt out my 'notorious' anti-Semitism. Apparently no one believes that I—and others—may also have something good and appreciative to say."[62]

At the same time that he was making this public claim of even-handedness, Jung was expressing more negative opinions in a letter to Wolfgang Kranefeldt.

As is known, one cannot do anything against stupidity, but in this instance the Aryan people can point out that with Freud and Adler, specific Jewish points of view are publicly preached and, as likewise can be proved, points of view that have an essentially corrosive character. If the proclamation of this Jewish gospel is agreeable to the government, then so be it. Otherwise, there is also the possibility that this would not be agreeable to the government...[63]

It was about this time that the first issue of the 1934 *Zentralblatt* appeared. Jung wrote the lead article "The State of Psychotherapy Today" (CW 10, pp. 157–173) and it has become the main source for those seeking conclusive written proof of Jung's anti-Semitism. Unfortunately, they take their quotes from just two paragraphs, ignoring the rest of the article and the specific context in which it came to be published. I will redress this in what follows, paying particular attention to the liberties taken by its translator that soften or eliminate some of Jung's more polemical phrases influenced by Nazi rhetoric.

As the new president of the International Society Jung's main goal was to insure the professional existence of psychotherapy under a regime that equated it with the "Jewish science" of Freudian psychoanalysis. Jung's willingness to accommodate himself is evident in the following statement "In Germany everything must be 'German' at present to survive. Even the healing art must be 'German,' and this for political reasons...It is a cheap jibe to ridicule 'Germanic [-racial] psychotherapy,' but it is very different thing to have to rescue medicine for humanity's sake from the seething chaos of revolution."[64] This was from his "Rejoinder to Dr. Bally" and the subject of a letter to Walter Cimbal, the Society's secretary. "You will appreciate that as its editor I must have some influence on its make-up at least in certain respects. You may rest assured that I will not under any circumstances use this influence for the publication of anything that is politically inadmissible."[65]

The title indicates that Jung intended the article to be his authoritative pronouncement about psychotherapy at a critical juncture in its history. It opens with Jung's lament about the "mechanization" and "pronounced ["ausgesprochenen," deleted in English] soullessness" of a psychotherapy that emphasizes technique over individuality. At first, Jung's comments were general in nature, criticizing unnamed schools for their "bigoted dogmatism and personal touchiness." He then elaborates on his long-standing views of Freud (who took his stand on sexuality with "fanatical one-sidedness") and Adler. The new variation on his old theme of their reductive orientation is that they "explain a neurosis from the infantile angle." For Jung the popularity of psychoanalysis was not so much due to its heuristic validity but "on the easy opportunity they afford of touching the other fellow on his sore spot, of deflating him and hoisting oneself into a superior position." Opportunism was apparently the least of psychoanalysis' faults since Jung then launched into what can only be described as a sarcastic diatribe against it. At least five times he dismisses psychoanalysis for its indulgence in "obscene fantasies" and also for its tendency to instill a "hostility to life" in its patients.

All this sounds worse in the original German. Hull translates "in den infantil-perversen Sumpf einer obszonen Witzpsychologie" ["into the infantile-perverse swamp of an obscene joke psychology"] as "to the level of a 'dirty joke' psychology" (par. 356). In the next paragraph, "to suspect their natural wholesomeness of unnatural obscenities is not only sinfully stupid but positively criminal" is more accurately translated as "to suspect their natural cleanliness [Reinlichkeit] of unnatural filth [Schmutz] . . . " Finally, Jung writes "die entwertende, zerfasernde, Unterminierungstechnik der <Psychoanalyse>" ["the depreciating, pulverizing, undermining technique of psychoanalysis"] but Hull conveniently deletes "pulverizing [zerfasernde]" (par. 360).

The most frequently quoted passages come from Jung's discussion of the contrast between Aryan and Jew (paragraphs 353–354). He begins with a reference to the "negative psychologies" of Freud and Adler and goes to explain that as Jews they had a special talent for discerning the shadow side of people. He then explains this in terms that mix insight and insensitivity: Jews have developed this talent because of their historical status as a marginal group ("this technique which has been forced on them through the centuries") but have aimed it "at the chinks in the armour of their adversary." He goes on to support his view with observations that he first made in his 1918 article "The

Role of the Unconscious." Jews, as members of a three thousand year old culture race, have a wider area of consciousness than Aryans; that as something of a nomad, the Jew lacks contact with the earth and needs a host nation for his development.

Jung repeated his old caveat that it was a grave mistake to apply "Jewish categories" to Christian Germans and Slavs. "The 'Aryan' [no quotation marks in the original] unconscious... contains explosive forces and ['creative' (deleted)] seeds of a future yet to be born, and these may not be devalued as nursery romanticism without psychic danger."

> Where was that unparalleled tension and energy while as yet no Nationalism Socialism existed? Deep in the Germanic psyche in a pit that is anything but a garbage-bin of unrealizable infantile wishes and unresolved family resentments. A movement that grips a whole nation must have matured in every individual as well. That is why I say that the Germanic unconscious contains tensions and potentialities which medical psychology must consider in its evaluation of the unconscious.

Jung criticized psychoanalysis on cultural-racial lines and implied that his approach was a viable alternative to it when he says that medical psychology must be broadened to include the "creative [constructive] powers of the psyche labouring at the future."

In the final paragraph of the article Jung put this whole issue in its historical context.

> It is the fate and misfortune of [medical] psychotherapy to have been born in an age of enlightenment... there is no sense in an entire generation of doctors to sleep on Freud's laurels. Much has still to be learnt about the psyche, and our especial need today is liberation from outworn ideas which have seriously restricted our view of the psyche as a whole.

Jung was aligning himself here with a group of German therapists trying to establish a position for their profession independent from its identification in the popular mind with psychoanalysis.

Jung's criticism of psychoanalysis was grounded in his broader critique of nineteenth-century materialism: "his [Freud's] materialistic bias in regard to the religious function of the psyche." His comments about the "obscenity" of psychoanalytic interpretations and his adoption of a cluster of Nazi buzzwords such as "pulverizing," "cleanliness/filth," and "constructive" stem from the deep-seated animosity that he still harbored toward his old psychoanalytic colleagues. In fact, he felt they, not he, were to blame for his problems when he said, "my own warning voice has for decades been suspected of anti-Semitism. This suspicion emanated from Freud. He did not understand the Germanic psyche any more than did his Germanic followers [*Nachbeter*—'parrots']."

Jung's Jewish Circle

Jung's article prompted an immediate reaction from a group of Jewish followers in Berlin that was drawn to him in the late 1920s (his June 1933 visit was occasioned by the formal founding of the Jung Society there). It elicited letters and a series of articles in the *Jüdische Rundschau*, the Zionist journal that continued to appear biweekly until 1938. This highlights a little-known fact of Nazi policy, namely, that in its early stage it permitted a surprising degree of freedom to Jewish publications within the strictly segregated cultural scene created by their exclusion from all other areas of German life.

These are important documents because in them his Jewish followers tried to come to terms with what Jung meant by "Jewish psychology." On May 26 Jung had written a letter to James Kirsch who had emigrated to Palestine only days after Hitler came to power.[66] He addressed only one point that he had made in his article, clarifying what he meant by his statement that the Jews had created no cultural form of their own. He acknowledged that conditions in Palestine may take things in a new direction but wrote that "the specific cultural achievement of the Jew is most clearly developed within a host culture, where he very frequently becomes its actual carrier or its promoter." It is important to understand that his positive evaluation of the Jewish experience is expressed only in a private letter to a Jew and did not appear in the original article. Another interesting shift is that Jung refers to the differences between "Jewish" and "Christian" rather than "Aryan" psychology and so avoids a discussion of his more troubling views on the Germanic psyche and National Socialism. This shift gives him an opportunity to express his feeling that Jewish hypersensitivity had led to "antichristian" attacks on himself. He cites as an example the *Israelitische Wochenblatt*'s contention that he had compared Jews to Mongolian hordes. Since the editors of the Letters note that the expression "Mongolian horde" had not been used, we have to conclude that the hypersensitivity involved here was, in fact, Jung's own.

He then identifies the source of this sensitivity. Both in private letter and public article he expressed the feeling that he had been victimized for twenty years by the accusations of anti-Semitism made against him by Freud "because I could not abide his soulless materialism."

> You ought to know me sufficiently well to realize that an unindividual stupidity like anti-Semitism cannot be laid at my door... [helping a person discover his

> individuality] is possible only if he acknowledges his peculiarity which has been forced on him by fate. No one who is a Jew can become a human being without *knowing* he is a Jew... (Emphasis in the original)

Jung was oblivious to the fact that, in light of events in Germany, his frequent and vocal opinions were liable to misuse and misunderstanding.

In a letter he wrote soon after to Gerhard Adler who was still in Berlin Jung continued this train of thought.[67]

> It is typically Jewish that Freud can forget his roots to such an extent. It is typically Jewish that the Jews can utterly forget that they are Jews despite the fact that they know that they are Jews. That is what is suspicious about Freud's attitude and not his materialistic, rationalistic view of the world alone... So when I criticize Freud's Jewishness I am not criticizing the Jews but rather that damnable capacity of the Jew, as exemplified by Freud, to deny his own nature... I speak in the interests of all Jews who want to find their way back to their own nature.

This is the most precise formulation Jung ever made of what he meant by "Jewish psychology." Ironically, it was the psychology of the assimilated Jew who had assumed a modern identity at the expense of his Jewish heritage. It is here that Jung clearly separated himself from the racial-biological anti-Semitism of the Nazis and where we find the reason for Jung's appeal to a group of Jewish analysts. It had to do with his commitment to helping people discover the roots of their individuality in the spiritual heritage of their given ethnic group. "This is the basis from which he can reach out to a higher humanity. This holds good for all nations and races. Nationalism—disagreeable as it is—is a *sine qua non*, but the individual must not remain stuck in it."[68] Rooted in the Romantic concept of the *Volksseele* (and to which he would return in his "Wotan" article), it was closely related to concerns expressed by the founders of Zionism. It is no surprise, then, to learn that this group of Jewish Jungian analysts had all been active in Zionist activities during their student years. Kirsch later reminisced that "Under the influence of the Zionist hiking club, the 'Blau-Weiss,' I became quite Zionistic myself."[69] Adler remembered that he and Neumann belonged to a student group that discussed the Jewish question along with other important topics.[70] Like Kirsch, Neumann emigrated to Palestine where he lived until his death in 1960.

What follows are the most relevant quotes from the *Rundschau* articles of the three men.[71] The first from Kirsch's "Some Remarks Concerning an Essay by C.G. Jung" in the feature "The Jewish Question in Psychotherapy" (May 28, No. 43, p. 11).

> In that Jung sees Freud in this way as a typical Jew, Jung comes to a conception of the Jews that is in fact characteristic of the Galuth [Exile] Psychology in

general and for the nineteenth century especially, but is surely not the last word about Jewish psychology. It seems that due to his decades-long fight with Freud only the Galuth image stayed stuck in Jung. He did not get past the phenotype of the Jew living in exile from the Schechina [Dwelling] to the genotype of the real Jew. In this way he oversees the real tragedy of Freud and the whole Galuth, namely that we've lost the connection to the creative depths of the soul. Jung comes therefore to mistaken judgments, for example "it is less dangerous for the Jew to put a negative value on the unconscious." On the contrary, it is especially characteristic but also especially dangerous for the Jew living in exile to destroy the connection to the unconscious. The "culture form" of the Jew—he has always had one of his own—is an especially unique form of dealing with the unconscious. Surely there has already developed here in old-new land a new type of Jew who accepts himself and his peculiarity and says yes to all the forces of life. But there is still the bigger task before us—to rediscover in the soul also the living connection with the elementary forces. In this respect the great psychologist Jung, who until now has been especially from the Jews treated with hostility and ignored in silence, can become a superb helper. Because exactly in Jung's personality, his psychology and psychotherapy, there is contained something that speaks to the sick Jewish soul in its depths and can lead to its liberation.

Erich Neumann sent a letter excerpted in the same feature on June 15 (No. 48, p. 5). He challenged Kirsch's contention that Jung had not gotten past his experience of "the phenotype of the Jew living in exile" saying that Jung had based himself on his work with Jews.

It is wrong to emphasize a "special connection of the Jew to the eternal primal depths," even if it actually may once have existed. Jung doesn't deny that the Jews of the Bible beheld and lived the "larger aspect of the human soul" but his work on contemporary Jewish people caused him to see the clear and fateful tendency to repress this larger aspect. That's what matters today. We believe that Jungian psychology will become decisive for the attempt of the Jews to come to their fundaments: especially the so-to-speak "Zionistic" character of his findings will be path-breaking. This is similar to the way Zionism includes the irrational of the creative human primal depths.

In his "Is Jung anti-Semitic?" (August 15, No. 62, p. 2), Gerhard Adler also took issue with Jung's claim that the Jews had not created a cultural form.

I admit Jung's formulation here is very short and condensed, but it all depends on the context. But besides that the burden of proof lies on us to prove the opposite! Where he [Jung] attacks the Jews he does this in so far as they are negative and uprooted. Is he anti-Semitic for this reason? And that is the reason why. Especially today Jung does not remain silent about the Jewish question. A person of the importance of Jung is not only concerned with the neurotic situation

> of a single people, but rather with these people as exponents of their time who are looking everywhere for their fundaments and roots.

These men sought to explain to a Jewish readership the essence of their teacher's views on Jewish psychology by demonstrating that it was not only *not* anti-Semitic, but were, in fact, compatible with a Zionist perspective. Prudence dictated that they not take up in public Jung's views on the German situation. Privately, at least one of them, James Kirsch, had tried to make him realize the grave danger posed by the Nazis.

> We differed very sharply in 1933, when he had forgotten his understanding of the German unconscious as he had described it in a paper written in 1918. I did not accept his advice that I stay in Germany, even though he said that the Nazi system would be over in six months. When I saw him for the first time years later, in 1947, after the Second World War, the first words were those of sincere apology for the advice he had given me in 1933. "Of course, you were right, not I," he said.[72]

This conversation would have taken place when Jung visited Berlin in June as the following recollection by Kirsch makes clear. "I had a talk with Jung in 1933 when I took him to the Anhalter Bahnhof in Berlin for his return to Zurich. At that time, he just did not believe me that the Nazis were as awful as they actually were, but it was only in 1937 on that he understood the nature of the Nazis."[73]

Jung had contacts at this time with other Jews as well. In his *The Reality of the Psyche*, the fourth in a series of collaborative anthologies entitled "Psychological Proceedings" that he had published periodically throughout his career, Jung included a contribution by Hugo Rosenthal "just to annoy the Nazis and all those who have decried me as an anti-Semite."[74] Jung's strategy of balancing contrary points of view (soliciting one piece from a Jew as well as another from Kranefeldt, his most polemical German follower) reflected a deeply personal trait rooted in a Swiss value system that placed a premium upon compromise. This strategy guided his actions in the turbulent years ahead but whose shortcomings were later captured in a reminiscence of Jolande Jacobi: "I always called him a Petainist... He always wanted not to get into difficulties with people."[75]

The balancing act is reflected in the mix of lecturers at the Eranos conference and the Psychology Club Zurich. William McGuire notes that among the Eranos speakers were Martin Buber and a group of German and Italian scholars who would be expelled from their academic positions but also Jacob Hauer, the founder of the German Faith Movement, and Richard Heyer, both of whom were to join the Nazi Party when membership opportunities were reopened in 1937.[76] Jung was tolerant of religious experimentation, especially by those involved in the German Faith Movement. Heyer had switched his declared religious affiliation from Protestant to *gottgläubig* ("believer in God").[77] In May, Wilhelm Laiblin, a former member, like Adolph von Weizsäcker, of Hauer's *Kommende Gemeinde*, gave a talk to the Psychology Club of Zurich entitled "The Struggle of Faith in Germany—Breakthrough or Breakup?" He later became an analytical psychologist in Stuttgart, which became a major center of Jungian psychology after the war.[78]

Jung's little-known relationship with a Jewish member of the Club Waldimir Rosenbaum sheds light on his maneuverings as the new president of the International General Medical Society for Psychotherapy as well as providing a poignant insight into Jung's conflicted personal feelings at the time. When Jung replaced Kretschmer as president of the General Medical Society for Psychotherapy he began to implement its reorganization as an international organization. This was necessary because of the creation in September 1933 of The German General Medical Society for Psychotherapy. This involved Jung in the intricacies and frustrations of administrative negotiations for the first time in twenty years. A new constitution had to be drafted for consideration at the Society's next conference to be held at Bad Nauheim from May 10 to 13, 1934. It was a busy place that month; three days after the psychotherapists vacated Bad Nauheim a secret meeting was held there in which the German military came to an understanding with the Nazi Party, a meeting that led directly to the murder of Ernst Röhm and other leaders of the SA the following month.[79]

Besides promoting the formation of national groups, Jung sought other ways to limit the influence of the overwhelming German membership. He solicited the advice of a young Jewish lawyer Wladimir Rosenbaum who was, along with his first wife Aline Valagnin, a member of the Psychology Club Zurich. Jung came to him with a draft of the Society's new bylaws.

> He [Jung] said he wanted to try to moderate these new by-laws; and by formulating them as little nazi-like as possible and in such a double meaning to make it possible to slip out of this whole Nazification... Jung took the changed draft and it was also accepted in the way I had revised it. Jung returned, and later came to see me to tell me about that. I remembered that very well. He came and said all the time, "They really are crazy!"[80]

Rosenbaum later became active in supporting the Republican side in the Spanish Civil War and spent four months in prison for violating Swiss neutrality laws. After receiving assurances that Jung would welcome his attendance at meetings of the Psychology Club, he went at a

> time several of the Club members were much influenced by the Nazis.—Then when I attended one of their evenings, several of the members obviously protested to Jung... And out of the blue, I received a letter from Jung: "I should

> not come anymore. Why, in the first place, had I attended?"...Then I asked Jung for a meeting together...I went there [Bollingen] at the appointed hour. Jung received me outside of the house; he didn't let me enter, but staying outside. I said, "Herr Professor, I came to ask you for clarification. I thought that I returned to the Club meetings with your approval. And now I got this letter. Please can you explain that to me?" Then Jung answered with a sentence which I won't forget for the rest of my life. He said to me, "Even a wild animal, being wounded, hides somewhere to die." And I remember that was for me like a blow. But, as a matter of fact, I reacted exactly as a boxer would. (I was a very good boxer; I was strong in taking it.) I just looked at Jung and said, "Goodbye, Herr Professor!"[81]

The bylaws of the new International Society that Jung lobbied for included two important provisions: one was that no national group could control more than 40 percent of the votes, the other provided for individual membership in the Society apart from membership in a national group. Although this was meant as a help to German-Jewish colleagues Cocks points out that Jews were not barred from membership in the German Society until 1938 when they lost their right to practice medicine.[82] There are several other things to consider about what Jung was dealing with in this matter. One is that the organizational model that Jung would have been familiar with was that of the Kulturbund, which was based on independent local chapters that cooperated in a series of annual conferences. Second is the fact that the German group was not merely a chapter of the international organization but was an autonomous society that existed independently of it. Nazi authorities were adamantly against any German organization being subordinate to international authority.

The German Society published *Deutsche Seelenheilkunde* ("German Psychotherapy"), eight of whose ten essays also appeared in the 1934 *Zentralblatt*. They included contributions by Heyer ("Polarity, a Fundamental Problem in the Being and Becoming of German Psychotherapy") and Häberlin ("The Importance of Ludwig Klages and Hans Prinzhorn for German Psychotherapy"). Heyer located Jung and Klages in the tradition of Romantic *Naturphilosophie* and praised their sensitivity to chthonic life, contrasting German polytheism with the monotheistic teaching of the "Jewish Yaweh-Spirit."[83] In reviewing the book in the October 15 issue of *Hippokrates*, Heyer noted that the ten authors were like an orchestra, playing their different instruments in the interest of a single motif: "the German soul in the new state." The review in the Nazi medical journal *Zeil und Weg*, however, concluded with the comment that "we do not recognize the ability of the Swiss C.G. Jung as an educator of the German Volk-Community."[84]

Although suspect in the eyes of some of the ideologues of the Nazi establishment, Jung was getting generally favorable press in Germany. Eugen Heun of Berlin, a member of the General Medical Society for Psychotherapy, wrote

"On the Collective Unconscious" for the January issue of the *Zeitschrift für Menschenkunde.* Jung's involvement in German cultural affairs led to new honors. On June 21, 1934, the first anniversary of his acceptance of the presidency of the Society, he was made a member of the Kaiserlich Leopold-Carolinisch Deutsche Akademie der Naturforscher, one of Germany's oldest and most prestigious scientific bodies (one of its former presidents was Carl Gustav Carus). One of the German honorees was Johannes Stark, a longtime Nazi and promoter of a "German physics" purged of the influences of the Jew Einstein. Other Germans included Fritz Lenz and Otmar Freiherr von Verschuer who became important figures in the world of Nazi science (von Verschuer later became Josef Mengele's mentor).[85] An analysis of the membership list does show that almost 50 percent of the new members were foreigners and indicates an effort by the Akademie to maintain its ties to the international scientific community during the process of *Gleichschaltung.*[86]

Jung's Other Publications

Jung's article for the *Zentralblatt* "The State of Psychotherapy Today" was the most important thing he wrote in 1934 and has long been controversial as his most polemical statement in the first years of the Nazi regime. Jung's opinion that psychotherapy had to be grounded in a nonmaterialistic philosophy was expressed in another piece he wrote that year. It was a foreword to *The Wonder of the Psyche* by Carl Ludwig Schleich (1859–1922) who made one great contribution to medical science, local anesthesia. This particular book, however, contained his later, more speculative ideas about the relationship of the sympathetic nervous system to the psyche and what he called the "World Soul." In his autobiography he wrote "All is movement, idea, flux. The universe has become completely spiritualized... Life is a manifestation of universal spirit."[87] It seems likely that Jung had heard about him years before from Oscar Schmitz who had dedicated a 1914 book of his to Schleich.

Although he had reservations about Schleich's wilder flights of fancy and his naïve conception of dreams Jung clearly felt him to be a kindred spirit. Both men refused to accept the predominant materialistic premises of the science of the day, an affinity best captured by their mutual antipathy to Emil Du Bois-Reymond. As a medical student Schleich had been tested by Du Bois-Reymond on the sympathetic nervous system and had been given a difficult time.[88] For his part, Jung wrote in a letter contemporaneous with the review that "he [Freud] is simply an exponent of the expiring 19th century, just like Haeckel, Du Bois-Reymond, or that *Kraft und Stoff* ass Büchner."[89]

Jung's argument criticizing the materialistic bias of modern science and supporting a more holistic approach was one he held consistently throughout his career. Readers familiar with his interests will also not be surprised that Jung gave the review some focus by comparing Schleich's work with that of one of his favorite figures, Paracelsus. He was drawn to Paracelsus' alchemical work and saw him as a forerunner of his psychological approach. By 1934, interest in Paracelsus extended beyond Jung and the small group of Paracelsus scholars. He

> appeared in Nazi books and magazines as the personification of German medical science. Paracelsean medicine was said to embody the natural, earthbound, experimental character of German medicine—medicine that was "close to the people" and not based on "a lot of complicated theories." It embraced "the whole man," not just particular organs or ailments.[90]

He became the patron saint of the Committee for a New German Science of Healing founded in 1935 that embraced most of the alternative health community and included the German General Medical Society for Psychotherapy.[91]

The review is important because it further documents the extent to which Jung was influenced by his audience and the atmosphere of the time. The most explicit appropriation of Nazi rhetoric comes when he writes that Schleich "fought shoulder to shoulder with me for the recognition of the soul as a factor *sui generis*..."[92] The phrase "fought shoulder to shoulder" was very popular in Germany at the time; it harkened back to the *fronterlebnis* (the experience of combat shared by veterans of World War I) that was held up by the Nazis as a model for the party and indeed the entire German nation. Jung also characterized both Paracelsus and Schleich as "revolutionaries," a reminder that "old fighters" were to be found on the cultural as well as the political front. Besides being revolutionaries he called them both "enthusiasts," which he understood in its original meaning as "god-filled' and is thus a passing reference to his view that Wotan the god of the Germans had been activated.

Another word that functions as a cultural "complex indicator" is "blood." "One of Schleich's favourite ideas was that of a psyche spread through the whole of the body and dependent more on the blood than on grey matter. This is a brilliant notion of incalculable import."[93] Several sentences later Jung mentions their shared interest in "the mysterious connections between the psyche and the geographic locality." Although he understood these terms differently from the Nazis, Jung's reference would have evoked the Nazi ideograms of *blut* and *boden* ("blood" and "soil"). (Jung's other explicit reference to blood was in his Radio Berlin interview where he said that the nobility believed in the "blood and in racial exclusivity.")

Finally, the foreword's closing sentence echoed the theme of "liberation" that concluded "The State of Psychotherapy Today" when it referred to the value of Schleich's work as a "liberation from the narrowness of mere academic specialization."[94] This situation stemmed from the materialistic philosophy that had become dominant during the Wilhelmine era, a period in which Germany made incredible industrial progress due in great part to its tremendous scientific achievements. Conservative intellectuals, however, felt that the country had paid too high a price for that progress and blamed the positivistic premises that had come to dominate academia since the mid-nineteenth century. Scientists such as Schleich and Dacque carried on in the *Naturphilosophie* tradition that had been in retreat but had not disappeared completely from the scene. Jung was sympathetic to this traditional opposition to the overemphasis on the intellect, which he said had "turned into a ravening beast."[95]

The Nazi allegiance to an ersatz Nietzschean *Lebensphilosophie* resulted in an assault not just on the intellect but on intellectuals as thousands were expelled, imprisoned, and exiled. Jung's indifference to this assault on life and liberty can best be seen in a review he wrote of Count Keyserling's most recent book *La Révolution mondiale et la responsabilité de l'Esprit*. Since his attendance at the 1927 School of Wisdom Conference Jung and Keyserling had stayed in touch. Jung delivered "Archaic Man" to the Jubilee Conference of the School of Wisdom in 1930; Prince Karl Anton Rohan also spoke at the conference, which was organized by Oscar Schmitz. Keyserling visited Jung in Zurich and wrote him many letters during a trip that resulted in his book about South America.

Keyserling had the time for such an extended trip since the School had closed on account of financial difficulties caused by the Depression. His personal situation become more tenuous after the Nazis came to power. He had criticized them in several articles in the *Kölnische Zeitung* in 1931–1932 as well as criticizing Alfred Rosenberg's *Myth of the 20th Century*: "Rosenberg's book made clear to me that National Socialism is, in its present form, basically hostile to the spirit."[96] Keyserling briefly lost his citizenship until the Prussian minister of the interior intervened and had his right to travel restored.

Keyserling was able to visit Paris in October 1933 for a meeting of the Permanent Committee of Arts and Letters of the League of Nations. Count Harry Kessler recorded these impressions of Keyserling in his diary. "He and Paul Valery conducted the whole congress, he had to make speeches the whole time, and everything went off splendidly. He has hopes of the alliance of a few hundred European intellectuals proving the salvation of European civilization. Lack of intellectuality is most terrible thing about the Hitler regime."[97] Keyserling's address was the basis for the book.

Jung opened the review with a discussion of the fact that Keyserling had written the book in French. He found it a "sign of the times" but did not state that the real reason why the book was published in Paris was because Keyserling could not find a publisher in Germany at that moment. The cultural subtext gets more interesting after we learn that Jung says that this situation is reminiscent of eighteenth-century Germany where the educated elite preferred French to their own "clumsy German." Linguistic nationalism was a heated topic in Germany in 1934 with the Nazi authorities promoting the use of Gothic script and the replacement of foreign words with their "authentic" German equivalents (e.g., in the 1935 *Zentralblatt* "psychotherapy" was replaced by "Seelenheilkunde" and "psychology" by "Seelenkunde"). Jung played to this prejudice when he said "I wish the book been written in German, for, in my unqualified opinion, its spirit is as un-French as it could possibly be."[98]

Jung then takes up his main argument, which is the critique of Keyserling's spiritual solution to the contemporary crisis. "How can that religious renewal predicted by Keyserling as necessary and imminent, come about unless our much vaunted spirit . . . can gracefully die? . . . What does the supremacy of the 'telluric powers' mean, except that the 'spirit' has once again grown weak with age, because it has been too much humanized?"[99] Jung was dismissive of what he considered Keyserling's optimistic, and now passé, Enlightenment perspective. The count's plan envisioned a "cultural monastery" that would produce a new spiritual elite ("Men whose consciousness is naturally centered on a plane superior to earthly happenings, to country, to race, to social and political necessities"). Jung made a counterproposal that reversed each of Keyserling's points: a true order would include men who, among other things, "have their natural centre of consciousness in their earth, their race, and in social and political necessities."

This list makes clear Jung's belief that a national agenda superseded an international one, a perspective he shared with Schleich who wrote that "Everything national is a blessing, everything international will sooner or later be a poison to the nation."[100] Jung, like many Swiss, had a low opinion of the League of Nations. He saw it as another example of the leveling process that was eliminating the vital differences that existed among nations as well as individuals.

Keyserling's son Arnold had this to say

> [Jung] considered the uprise [*sic*] of National Socialism as a rejuvenation . . . he wrote a criticism of my father's book, *La Révolution mondiale et la responsabilité de l'Esprit* saying that those people who consider themselves spiritual leaders should firstly get in touch with the forces of instinct because it's out of these forces of instinct that also the spiritual rejuvenation can only come. So that was taken as an attack on my father, because my father was not clearly labeled up to that moment as to what he really taught, but now they knew it. He was

> against—which was quite true—the forces of blood, race and so on because my father considered National Socialism as the first really anti-spiritual movement in European history and much more dangerous than anything else. Jung also thought the same thing but thought it a necessary evil, and that was the trouble between them. What my father didn't like in this thing was that Jung, in Switzerland, didn't consider the possibilities of Gestapo and things like that. It is quite normal, how should a Swiss citizen be aware of things of that kind? And that was the trouble. But it should never be said that Jung was a Nationalist Socialist.[101]

Jung's review showed no sympathy for an old friend who was now vulnerable to the whims of Nazi authorities. In this and in his Schleich piece Jung aligned his argument to the views of his German colleagues who actively championed the biocentric philosophy of Ludwig Klages.

Remembering that Ringer talked about a "constellation of attitudes and emotions" that infect language, we can see that during this period Jung accommodated his language to that of the times. Besides the egregious concessions to Nazi rhetoric he relied on such abstractions as "the logic of history" and "destiny" to carry his arguments forward. The linguistic concessions were accompanied by a moral concession that rationalized that "sometimes the better first appears in evil form."

Chapter 5

Nazi Germany and Abroad

In many ways the mid-1930s marked the pinnacle of Jung's professional career. His sixtieth birthday in 1935 was celebrated by his followers with the publication of a *Festschrift* that assessed his contributions to psychology and the human sciences.[1] His stature in his chosen field had earned him the presidency of the International General Medical Society for Psychotherapy, a professorship at Switzerland's Federal Technical University (the ETH), and honorary degrees from such universities as Harvard (1936), Yale (1937), and Oxford (1938). This recognition led to his popularity as a commentator on the increasingly unsettled world situation. Jung's fame brought controversy as well as kudos. In particular his attendance at the Harvard Tercentenary was criticized by those who found his statements about Aryan and Semitic psychologies offensive. Their charges were to be repeated with even greater publicity after the war. What has not been known is the extent to which Jung's work was favorably reviewed in the Nazi-controlled press. In this chapter we will follow Jung as he navigates the increasingly turbulent international waters.

The International General Medical Society for Psychotherapy

After the reorganization of the Society at Bad Nauheim in 1934 Jung assumed his new responsibilities as president, his first goal being the formation of the Society's constituent national groups. He was able to announce in a 1935 *Zentralblatt* editorial that there were now functioning groups in Holland, Denmark, and Switzerland while delays were being encountered in Sweden.[2]

His administrative responsibilities necessitated correspondence with the heads of the other groups: J.H. van der Hoop (Holland), Oluf Brüel (Denmark), and Poul Bjerre (Sweden).

J.H. van der Hoop was a Dutch psychotherapist who had published his *Character: The Unconscious, a Critical Exposition of the Psychology of Freud and Jung* in 1923. Jung would have been pleased with its even-handed treatment of the Zurich School's synthetic approach to psychology and its emphasis on the cooperation between the conscious and the unconscious. Later, in 1940 Jung tendered his resignation as president but not before considering van der Hoop as his successor.[3] Oluf Brüel, like Jung, was a contributor to William McDougall's journal *Character and Personality* where he reported on news from the International General Medical Society for Psychotherapy. The concern for the Nordic psyche that he expressed in articles of this period indicate a conservative stance that was in general agreement with McDougall and his German colleagues. The best-known of the three was Poul Bjerre. He had attended the 1911 Psychoanalytic Congress at Weimar in the company of his lover Lou Andreas-Salome but eventually sided with Jung after his split with Freud.[4] After this there was a long hiatus until they became reacquainted through the International Society. Bjerre was the head of the Swedish group that formed in 1936 and proposed "Race and Depth Psychology" as the theme of that year's annual congress.

The Dutch group had initially offered to host that Congress as a way of escaping the politicized climate of Germany that was in evidence when the Society had met again at Bad Nauheim in 1935. Apparently many members of the group still had reservations about this and the invitation was rescinded.[5] Jung's letter to the group showed the lengths that he had to go to in order to balance the competing political views within the Society. He reminded them that "our German colleagues were not the makers of the Nazi revolution, but live in a State that demands a definite political attitude." He implicitly criticized the Dutch for their leftist stance when he said "I am convinced that if Russian doctors who believe in the religion of Communism sought to join the International Society the present opposition would raise no objections."

Although no congress was held that year, efforts to hold one outside Germany were finally successful when the Society met in Copenhagen from October 2 to 4, 1937. By then an Austrian group had been established (which was to be absorbed into the German group after the 1938 annexation) and creation of an English group was approved. Jung's lectures at the Tavistock Clinic in 1935 and 1936 had generated interest and fostered many professional contacts in London. The English group sponsored the last congress held at Oxford in 1938. Göring, leader of the German group, was unhappy with these developments and unsuccessfully opposed the approval of Erich Strauss, a Jew, to a leadership position in the English group. He then lobbied Jung to approve the admission of pro-Nazi groups from Hungary, Italy, and Japan; Jung found this unacceptable and with the outbreak of the war resigned his presidency in 1940.

Developments in the *Zentralblatt*

When Jung took on the presidency of the General Society, he also took on the editorship of its official journal—the *Zentralblatt*—which continued to be published in Germany by S. Hirzel of Leipzig. Jung, whose distaste for organizational details was well known, did not edit manuscripts, and in fact, according to the testimony of the Society's general secretary C.A. Meier, left the dirty work of rejecting many anti-Semitic articles to Meier.[6] Jung did contribute several editorial pieces to the 1935 issue, reiterating views he had held almost unchanged since his break with Freud. What was new was that they were now expressed by Jung in his capacity as president of an international society—and in the context of Hitler's increasingly persecutory regime.

In the first of the *Zentralblatt* editorials, Jung elaborated on the necessity for psychotherapy to broaden its narrow concern for case histories to encompass a regard for a person's *Weltanschauung*. He went on to discuss the importance in psychotherapy of fostering a normally adapted attitude.

> [Adaptedness] is a continually advancing process which has as its indispensable premise the constant observation of changes occurring within and without. A system of healing that fails to take account of the epoch-making *représentations collectives* of a political, economic, philosophical, or religious nature, or assiduously refuses to recognize them as actual forces, hardly deserves the name of therapy. It is more a deviation into a pathologically [morbid] exaggerated attitude of protest which is the very reverse of adapted.[7]

This is a reprise of the argument he made in his "Rejoinder to Bally" where he said "[Science and every healing art] must learn to adapt themselves. To protest is ridiculous—how protest an avalanche? It is better to look out."[8] This is Jung's accommodationist rationale in a nutshell, but with a troubling new qualification. Whereas earlier, protest was only ridiculous, it was now labeled "morbid." In light of the consolidation of Nazi control over Germany, Jung's statement would have been understood by its readers as support, however unwitting, for the regime's efforts to silence its critics.

In his presidential address Jung first discussed one of the Society's main concerns, namely, the right of medical psychotherapy to a professional existence independent of psychiatry and neurology. (A great deal of the Society's organizational maneuverings was due to this concern.) Jung then turned to the differences that divided the psychotherapeutic profession. Without mentioning them by name, Jung took his former psychoanalytic colleagues to task. "There are certain groups of doctors who put forward theories with totalitarian pretensions and barricade themselves against criticism to such an extent that their scientific convictions are more like a confession... these psychological theories are notably intellectualistic as well as anti-religious."[9] The animosity that he harbored toward his Freudian critics, evident in "The Present State of Psychotherapy," continued to blind Jung to the true nature of contemporary totalitarianism. He consistently solicited sympathy for his German colleagues and would argue in his 1936 article "Wotan" that the Germans should be seen as victims of that archetype.

The formation of the various national groups in the Society prompted Jung to sponsor special issues for each, the articles from which were incorporated into the general edition. In his 1935 editorial note Jung announced that following a Scandinavian and a Dutch issue it was now Switzerland's turn. "Just as there are points of view based on race psychology, so also there are national ones, and we may welcome it as an enrichment of our experience that we succeeded in including in our issues contributions from the Romance and the Anglo-Saxon mind."[10] Surprisingly this special Swiss issue had only one contribution besides Jung's from the fourteen-member Swiss Society for Practical Psychology. There were, however, two other Swiss contributors. One was J.B. Lang, a colleague from Jung's early years who is most famous as Herman Hesse's analyst and the inspiration for Pistorius in Hesse's *Demian*. The other was Charles Baudouin who had drawn closer to Jung after attending his 1934 Basel seminar. The two "Anglo-Saxon" contributors were Jung's longtime students H.G. Baynes and Esther Harding.

Eleven of the thirteen other articles that appeared in the 1935 general volume were papers that had been delivered at the Bad Nauheim Congress that year. Most were technical in nature and warrant little comment today but another indicates the frequent and favorable reference to Jung evident in the journal during this period. It was entitled "The Collective Unconscious of C.G. Jung, its Relationship to Personality and the Group Soul" and was written by Otto Curtius who had been moving closer to Jung and was eventually made a guest-member of the Psychology Club Zurich. At the same time, he was assuming important administrative positions with the *Zentralblatt* and the German General Medical Society for Psychotherapy. Something that he wrote in honor of Jung's sixtieth birthday reflects the esteem in which he and his German colleagues held Jung. "From the deep layer of the Nordic-scientific tradition... has Jung conceived his fundamental ideas and erected the edifice of his scientific teaching."[11]

Establishing the pedigree of this tradition had begun, as we have seen, among the intellectual opponents of the Weimar republic. After 1933 their views were officially sanctioned and promoted by the Nazi cultural-scientific authorities. A good example of this kind of writing was Carl Häberlin's article in that same issue. He predictably relied on Klages' dichotomy of logocentric/biocentric to critique the mechanistic view dominant in the nineteenth century and to offer an authentic German alternative ("German" was constantly being used to legitimize things). His pantheon of great German "soul-researchers" (with Heraclitus as a spiritual ancestor) includes Paracelsus, Goethe, Carus, and Nietzsche.[12] He announced that they were all deep thinkers who shared a holistic vision that was uniquely German. This was all standard stuff for him but two further comments reveal his accommodation to the new intellectual guidelines. At one point he mentions the difference between "Semitic and our thinking"; the other was a concluding remark affirming the role of German psychotherapy in developing a genuine community under the auspices of the National Socialist state.

The format of the journal was to remain unchanged from that of its early years. Besides the articles there were various organizational announcements, congress schedules, and reports. One of the most important features of the journal was its extensive book review section. Of interest are the changes that did take place in that section after Matthias Göring became Jung's coeditor in 1936. After that year separate sections on psychoanalysis and individual psychology were dropped, although books about both continued to appear in the "Depth Psychology" portion of the "Psychotherapy" section. Reflecting an effort to broaden the narrow technical orientation of the field, a "Philosophy" section was added. Two other new sections reflect concessions to the reigning scientific agenda: "Hereditary Biology and Racial Science" and "Folk Psychology."[13]

That Göring became coeditor in 1936 reflects important institutional developments in the psychotherapy movement in Germany. In that year the Reich Interior Ministry established the German Institute for Psychological Research and Psychotherapy. The Institute consolidated the German General Medical Society for Psychotherapy, the German Psychoanalytic Society, the C.G. Jung Society, and Künkel's Work Group for Applied Character Studies. Its membership included most of the psychotherapists who had remained in Germany. Establishing branches throughout the country it had an elaborate training program and served the needs of other institutions (it eventually received funding from, among others, the Labor Front and the Luftwaffe). As we shall see, Jungians played a prominent role in the life of the Institute.

Göring's article "Weltanschauung and Psychotherapy" opens with a long quotation from Hitler about the importance of developing a common *Weltanschauung* for the folk community. Göring went on to explain why it was important for psychotherapists to participate in this process. "We must also study the mental life of our folk community, so we can have a picture of the psyche of our people. I say explicitly our German folk since I am of the opinion that within the Aryan race, every folk has its own mental peculiarities."[14] He then compared Freud's and Jung's understanding of libido and declared that the difference was rooted in their contrasting Jewish and Aryan *Weltanschauung*. Emphasizing their incompatibility, he recalled that Jung had acknowledged this long before 1933 and noted the segregation of the races with regard to their psyches. The article concludes with the stirring declaration "Every race, every folk should seek its own psyche."

Göring also referred to Jung in a review he wrote of *The Nordic Soul: an Introduction to Racial Science* by Ludwig Clauss, one of the most prolific "experts" in that popular new field. Unlike other such experts who emphasized physical characteristics, Clauss' talent was for elucidating the inner relationship between the various races and their native landscapes. Göring noted that this was similar to the type theories of Jung and Kretschmer.[15] There is no evidence that Jung found any of this objectionable. That he was aware of what Göring was publishing at the time can be seen in a letter he wrote to Göring in November 1937.

> Dr. Meier has drawn my attention to your short review of Rosenberg's book [*The Myth of the XXth Century*]. For anyone who knows Jewish history, and in particular Hassidim, Rosenberg's assertion that the Jews despise mysticism is a highly regrettable error. I would therefore suggest that we pass over this book in silence. I cannot allow my name to be associated with such lapses.[16]

Jung confined his criticism to a narrow point of scholarship while passing up an opportunity to expose one of the major intellectual frauds perpetrated by the Nazis. He apparently did not find it necessary to challenge Göring's use of his ideas since they were consistent with views he had publicly expressed since 1918. He either did not realize or did not care about the extent to which his work was to be utilized to provide intellectual legitimacy for the Nazi ideological agenda.

Jung's Followers and the Göring Institute

The Göring Institute, as the German Institute was popularly known, incorporated psychoanalysts, Adlerians, Jungians, and independents on its staff. Among the Jungians were Gustav Richard Heyer, Wolfgang Kranefeldt, Adolf von Weizsäcker, and Olga Koenig-Fachsenfeld. Koenig-Fachsenfeld had participated in the first German-language seminar held in Zurich in 1930. In 1935 her dissertation "Transformations of the Dream Problem from the Romantics to the Present" was published with a foreword by Jung.[17] She had first been introduced to Jungian thought while in Munich as a student of Heyer.

Before Heyer got deeply involved in administrative positions at the Institute he had continued his involvement with Jungian activities in Switzerland. He was a member of the Psychology Club Zurich and contributed an article to the *Festschrift* published in honor of Jung's sixtieth birthday. The book was entitled *The Cultural Meaning of Complex Psychology* and was edited by Emil Medtner.[18] Heyer's article was "Institutions as an Ordering Principle, a Psychological Inquiry." He mentions "our Munich friend E. Dacque" while discussing the formative influence of ritual on early mankind. His analysis goes on to compare medieval and modern society, favoring the organic complexity of the former to "the inevitable grey uniformity of liberal creation." Heyer also spoke at the first three Eranos conferences (1933–1935), speaking each time in the favored position immediately after Jung. Another example of Jung's high regard for him is the fact that Jung reviewed two of his books.[19]

Another contributor to the Jung *Festschrift* was Friedrich Seifert (1891–1963), a professor of philosophy at the Technical University in Munich who had been introduced to Jungian psychology by Heyer. His paper "Idea Dialectics and Life Dialectics" compared Jung's ideas with those of Hegel and earned him a letter of thanks from Jung.

> It was always my view that Hegel was a psychologist *manqué*, in much the same way that I am a philosopher *manqué*. As to what is "authentic," that seems to be decided by the spirit of the age. Or perhaps the decisive factor is the historical development of the functions, as I have always suspected, but whose history would have to be written by a professional philosopher. This development is a very complicated affair, since it would have to be treated not in terms of the contents that have remained more or less the same in the history of civilization but in terms of form.[20]

Besides his contribution Seifert also reviewed the *Festschrift* several times. In the August issue of the *Neue Schweizer Rundschau* he chose to evaluate Jung's unique role in contemporary thought rather than merely summarizing its articles. "Jung's concepts do not stand in a simple relationship to the contemporary spirit." Dismissing the causal/reductive theories of Freud and Adler as passé, he pointed out that Jung's conception of the unconscious was in the tradition pioneered by Goethe and the Romantics that emphasized its positive, creative character. As such, it acted as a determined argument against the chief axioms of rationalism and individualism that as products of the Enlightenment had cut man off from his religious sensibilities. It was here that Jung made his great contribution with his attention to symbols and the intuitive function.[21]

Seifert and Heyer collaborated on a two-volume anthology *Der Reich der Seele* that was published by Lehmanns Verlag in 1937.[22] Seifert explored the relationship between Jung's ideas and those of Heidegger, asserting that Jung's concept of the "objective psyche" was equivalent to Heidegger's concept of existential totality. Although Heyer's article dealt with practical problems of body-work his references to the mind-body polarity revealed his indebtedness to Klages. The other articles, mostly by female colleagues, including Heyer's wife Lucy, dealt with dreams, child psychology, and the relationship between astrological symbols and the hexagrams of the *I Ching*.

In September 1937 Jung was in Berlin to give a two day seminar on archetypes at the Göring Institute. This unpublished seminar became the basis for "Concerning Mandala Symbolism."[23] After reviewing his concept of the collective unconscious, Jung analyzed a series of fifty pictures. Along with examples from patients and from Eastern religions there were examples from alchemy, an area of interest that was beginning to deepen for Jung. An anecdote about the seminar exists in two different versions. It concerns what Jung said as a column of German soldiers marched past the open window of the Institute. In one anecdote Jung said "there go the archetypes down the street."[24] In the other the loud singing of the Nazi troops prompted Jung to stop "to let this Nazi noise pass."[25] The first is more in keeping with Jung's neutral-ironic style since the other, written by a postwar defender of Jung, would have caused offense, which was something that Jung tried to avoid.

Mussolini was in Berlin at the same time on a state visit and Jung attended a parade in his honor. He recalled the experience a year later in his interview with the American journalist H.R. Knickerbocker.[26] The fact that he was only a few feet from the two dictators would indicate that he himself was there as an official guest of Matthias Göring and the Institute. Jung paid particular attention to the contrasting body languages expressed by the two men. Jung thought that Mussolini acted like a boy at a circus when the goose-stepping soldiers marched past and was so enthralled that he introduced it when he returned to Rome. "I couldn't help liking Mussolini. His bodily energy and elasticity are warm, human, and contagious." Hitler made an entirely different impression on Jung. "During the whole performance he never laughed; it was though he was in a bad humor, sulking. He showed no human sign. His expression was that of an inhumanly single-minded purposiveness..." Jung's observations of the two men are another example of his reliance on *Menschenkenntnis,* the intuitive method of evaluating character based on immediate visceral reactions. It leaves much to be desired since it stays on a superficial level. In this case Jung did not take into account other facts such as the impact of Mussolini's "energy and elasticity" on untold thousands of Italians and Ethiopians.

Other Activities in and Views of Nazi Germany

Jung's presidency of the International Society is his most well-known and controversial connection to Nazi Germany. His reputation, ambition, and psychological interests led to a number of other activities that also deserve our attention. Besides giving a more complete picture of Jung's involvement they illustrate the *modus vivendi* he established with developments there. The articles he was to publish and the conferences he was to attend were uncontroversial in their content. In fact, it is their very "normalcy" that is significant since they raise questions about the moral choices people make vis-à-vis immoral situations. After the dramatic changes of 1933 and 1934 the "German Revolution" had entered a phase in which the government sought to institutionalize its vision of a Nazi state. Along with its internal agenda the Nazis sought to improve Germany's international reputation on a number of fronts, the most famous of which was its hosting of the 1936 Berlin Olympics.

The first article was "Psychological Typology" and it appeared in the February 1936 issue of the *Süddeutsche Monatshefte.* Most of it is a popular exposition of Jung's most famous contribution to scientific psychology and is similar to articles that had appeared in 1923 and in 1931. The only significant difference comes in the beginning where Jung gives the theory's historical background and refers to the materialistic presumptions of contemporary psychology. He goes on to criticize Freud who in keeping with the spirit of the age "narrowed the picture of man to the wholeness of an essentially 'bourgeois' collective person, and this led necessarily to philosophically one-sided interpretations [more accurately, 'to a one-sided *Weltanschauung* interpretation']."[27]

Although criticizing Freud for his materialistic bias was typical for Jung, its particular formulation here is of interest given the article's context. It appeared in an anthology *Moderne Seelenkunde* ["Modern Psychology"] that included contributions by Friedrich Seifert, Ludwig Klages, Gustav Richard Heyer, Matthias Göring, and Fritz Künkel. Jung's portrayal of Freud as the representative of a now outmoded nineteenth-century bourgeois worldview complemented that of the other contributors. Heyer developed this theme when he catalogued the shortcomings of this worldview with its undue emphasis on an individualism grounded in materialism and democracy. Göring's remarks had a more polemical edge and included his obligatory reference to Hitler's *Mein Kampf.* He recalled how the Munich group under Seif with their "Aryan instincts" had opposed the Marxist direction of the Jewish leadership of the Society of Individual Psychology. Furthermore, he went on to associate Jung's theory of the collective unconscious with the research of genetic biologists into the influence of ancestors on the human genotype.

Jung continued his critique of Freud in his English-language commentary on *The Tibetan Book of the Great Liberation.* There he wrote "Introversion is felt here [in the West] as something abnormal, morbid, or otherwise objectionable. Freud identifies it with an autoerotic, 'narcissistic' attitude of mind. He shares his negative position with the National Socialist philosophy of modern Germany."[28] In 1939, the year of Freud's death, Jung now added insult to injury by linking Freud's name with the very group that hated him and forced his emigration from his lifelong home in Vienna.

The other article by Jung that appeared in a German journal at this time involved his other contribution to scientific psychology, the word-association experiment. His expert opinion had been solicited by Zurich authorities in a well-publicized murder case and he analyzed the responses of the suspect. The results were published in the *Archiv für Kriminologie* (Leipzig) in 1937.[29] It had been founded by Hans Gross, Otto Gross' father, and Robert Sommer, one of the founders of the General Medical Society for Psychotherapy. It was a leading journal in scientific criminology (other articles dealt with hair analysis, counterfeit banknotes, firearms, and arson) with an international reputation (other authors were from the United States, Australia, and Sweden). Again, it is the context rather than the content of Jung's article that is troubling. If the journal itself remained relatively unchanged, this was only due to the Nazi government's desire to maintain the semblance of a press untainted by overt Nazi rhetoric. As many observers recognized, German criminology now relied on methods far less subtle than the word-association test.

Jung's ongoing interest in the relationship of psychology and religion also continued during these years. He and Bishop Stählin were the featured speakers at the annual conference of the Kongener Kreis at Königsfeld, Germany, in January 1937. The Kongener Kreis had been founded by Jacob Hauer in the 1920s but was currently headed by Rudolph Daur who would continue his connection to Jung after the war through his involvement with Wilhelm Bitter's Stuttgart Jungian group. The group was composed of those members uncomfortable with following Hauer into the German Faith Movement. Jung had most likely been invited through the efforts of Adolf Weizsäcker, his Radio Berlin interviewer and a member of both the Kongener Kreis and Stählin's Berneuchner Circle.[30] The theme of the conference was "Psychology and Spiritual Leadership" and its report entitled "On the Threshold" had contributions by Daur, Stählin, Weizsäcker, and Dr. M. Bircher-Benner, an old colleague of Jung's. Although Jung's talk apparently went unrecorded, it was probably similar to several recent papers that he had given to pastoral audiences, "Psychotherapists or the Clergy" and "Psychoanalysis and the Cure of Souls." In them he pointed out the inadequacy of Freudian and Adlerian theories in dealing with spiritual problems since they could not help people find meaning in life. This could only come about through a direct experience of the psyche, like that of St. Paul on the road to Damascus. In the first paper, a 1932 address to the Alsatian Pastoral Conference, he returned to his theme of modern and pseudo-modern man.

> I have found that modern man has an ineradicable aversion to traditional opinions and inherited truths. He is a Bolshevik for whom all the spiritual standards and forms of the past have somehow lost their validity, and who therefore wants to experiment with his mind as the Bolshevik experiments with economics.[31]

Jung's major article of this period was "Wotan," which appeared in the March 1936 issue of the *Neue Schweizer Rundschau*. It is the most referred to and quoted of Jung's pronouncements on Nazi Germany. Unfortunately, the quotes are taken from the translation of R.F.C. Hull in the Collected Works, which is less accurate than the first English translation by Barbara Hannah in *Essays on Contemporary Events*.[32] The most significant mistranslation occurs where Jung is commenting on Hitler's influence. "one man who is obviously 'possessed,' has infected a whole nation to such an extent that everything is set in motion and has started rolling on its course towards perdition."[33] A more accurate translation is "one man who is obviously possessed has possessed a whole people to such an extent that everything has been set in motion and has started rolling, and is thus inevitably embarked on a dangerous course." The first point is that Hull adopted Hannah's use of "infected," which conveys a medical metaphor that is not found in the original. At this point Jung was interpreting Germany in terms of religious not medical phenomenology. Inspired by Otto's concept of numinosum he distinguished between *Ergreifer* ("one who seizes") and *Ergriffener* ("one who is seized").

More problematic is Hull's substitution of "perdition" for "dangerous course." This implies an element of moral judgment that is not found in the original since something "dangerous" is not necessarily "evil." This mistranslation has had serious consequences since it has led many commentators to incorrectly conclude that by 1936 Jung had formed a more critical opinion of Nazi Germany than he had previously held.[34]

Jung's analysis in "Wotan" is based on what he had written back in 1923 to O.A.H. Schmitz about the need of Germans to have a new experience of God through a confrontation with their primitive side. Picking up on his earlier comment about early Germanic religion he said "Wotan disappeared when his oaks fell and reappeared when the Christian God proved too weak to save his Christians from fratricidal carnage."[35] Jung then went on to make the case that the Wotan archetype was an "excellent" hypothesis for explaining National Socialism and better than any economic, political, or psychological theories. (In the CW the hypothesis merely "hits the mark.") Wotan is the *Ergreifer* and the German people are the *Ergriffener*. For Jung, what Germany was going through was essentially a collective religious experience since people there were in a state of "enthusiasm" (*en-theos*: "the in-dwelling of the god").

After relating Wotan to various Greek deities, Jung discussed the importance of mythology for understanding the deeper dimension of the psyche. He said that

> it is a question of basic types or images which are inherent in the unconscious of many nations. The behaviour of the nation takes on its specific character from its underlying images, and therefore we may speak of an archetype "Wotan." As an autonomous psychic factor, Wotan produces effects in the collective life of the people and thus reveals his own character. For Wotan has a peculiar biology of his own, quite apart from the nature of man.[36]

This passage contains Jung's only references to the biology of archetypes and comes closest to aligning the archetypes with the *Volksseele* concept of his Romantic predecessors. Notably, and most worryingly, Jung stressed that Wotan was not only inherent in the deep levels of the German unconscious but implied that he somehow connected to the germ plasm of the German Volk. In fact, the Wotanistic elements that Jung identified in Nazi culture (blood and soil, archaic folk customs, the Aryan Christ, and the Nordic origin of civilization) were not so much products of the psyche as they were the product of a century of sustained ideological elaboration. For example, already in the mid-nineteenth century Wolfgang Menzel (1799–1873) wrote in his *Deutsche Mythologie* that Odin [Wotan] was the personification of that driving force of the German people that made them supreme in world history. Many German intellectuals such as Paul de Lagarde were drawn to an anti-rationalist, anti-democratic ideology that emphasized "a unique Germanic-pagan prehistory that was broken by Roman and Christian influences."[37] After the Bayreuth Festival became a pilgrimage site, Wagner's blend of German mythology, anti-Semitism, and chauvinism reached an international audience.

Jung's interpretation of Wotan was based on the work of two scholars, Martin Ninck and Jacob Hauer, both of whom he knew personally and discussed in his article. What he emphasized about them was what he had said several years before about Carl Ludwig Schleich and Paracelsus, namely that they were "enthusiastic" scholars. Ninck was a Swiss who had recently written a book *Wodan und germanischer Schicksalglaube* [*Wotan and Germanic Beliefs in Destiny*] (Jena: Diederichs, 1935). "It is, indeed, very objective and does full justice to the rights of science... One feels that the author is vitally interested in his material, and that the chord of Wotan is also vibrating in him. This is no criticism; it is the highest merit of the book..."[38] Ninck had written articles about Klages for the *Zeitschrift für Menschenkunde* and a series of books about mythology and the Romantics. Jung's interest in Ninck's work led to his being invited to speak to the Psychology Club Zurich on German mythology in 1937–1938 and on Celtic religion in 1942.[39]

Worldviews figured in the book's concluding chapter "Outlook" where Ninck contrasted the Nordic attitude toward belief with the ancient and Christian-Roman view. He then went on to note that Wotan had most recently been incarnated in the figure of Faust. Jung was so impressed with this that he wrote "The author's *Ergreiffenheit* has added life to programme, as is particularly evident in the last chapter (*Ausblicke*)."[40] Since Hull chose to delete this sentence one might assume that he did so in order to minimize Jung's sympathy for Klages' biocentric philosophy.

The enthusiasm that Jung appreciated in Ninck's scholarship was also something that he was experiencing in himself at this time. In his 1935 Tavistock lectures in London, Jung had talked about how he was personally influenced by the Wotan archetype when he was in Germany ("I know it has to be as it is. One cannot resist it.").[41] This candid confession indicates the extent to which Jung himself was affected. He went so far as to say "the worshippers of Wotan, in spite of their eccentricity and crankiness, seem to have judged the empirical facts more correctly than the worshippers of reason."[42]

The article shows just how closely Jung had been following religious developments in Nazi Germany. The Catholic and Protestant denominations had generally adjusted to the new regime with a minimum of soul-searching. The situation did lead to the appearance of the German Christians, a group that sought to purge Christianity of all traces of Judaism by rejecting the Old Testament and denying Jesus' Jewish ancestry.[43] Where the German Christians saw an opportunity, other Protestants saw a threat and formed a network known as the Confessing Church; the latter became the organization most dedicated to fighting state control of church affairs. Its most famous member was Dietrich Bonhoeffer who was later to die in a concentration camp. Although Jung dismissed the German Christians as a contradiction in terms he asked in a sarcastic footnote: "Is the *Bekenntniskirche* [Confessing Church] inclined to be equally tolerant, and to preach about Christ shedding His blood for the salvation of mankind, as did Siegfried, Baldur, and Odin among others?" (*Essays*, p. 14) Religiously inclined Germans, he suggested, ought instead to join the German Faith Movement headed by his friend Jacob Hauer.

> One cannot help being touched when reading Hauer's book [*Deutsche Gottschau: Grundzüge eines deutschen Glaubens—German Vision of God: Basic Elements of a German Faith*] if one regards it as the tragic and really heroic attempt of a conscientious scholar. Hauer was not aware of what was happening to him, but as a German [more accurately, "as a member of the German Volk"], he was called and moved by the inaudible voice of the *Ergreifer*. (Ibid.)

The German Faith Movement was founded at a convention held at Eisenach on July 29–30, 1933. It was an amalgamation of a half dozen smaller organizations that had been struggling to establish a völkisch religion on equal footing with the Catholic and Protestant churches. Hauer was chosen as the leader of the movement assisted by a committee that included Count Ernst zu Reventlow and Hans Günther. Reventlow was as reactionary as his sister Fanny was revolutionary. He was born in 1869 and was active in racialist party politics, founding the German Racial Liberation Party, one of the Nazi Party's early competitors. In 1927 he went over to the Nazis and became one of their Reichstag delegates. "He viewed it [the Jewish Question] as a multifaceted spiritual problem whose historical, social, and above all religious aspects had vastly more significance than the matter of biological heritage."[44] Hans Günther had been Germany's leading expert on racial science since the 1920s. Another member of the committee was Herman Wirth whose writings on a Nordic Atlantis appealed to Alfred Rosenberg and led to his involvement in Himmler's *Ahnenerbe* [Ancestral Heritage] at the same time. Members of the Movement pledged that they were free of Jewish or colored blood as well as not having any Freemason or Jesuit affiliations.[45]

Jung's interpretation was strongly influenced by what he had heard from Hauer ("I think I have honestly taken pains to understand the German phenomenon from the outside, at least so far as this is possible for anyone who has experienced the same thing though in a quite different way."[46]). The two men shared a common frame of reference in the ideas of Rudolf Otto with whom Hauer had collaborated in the 1920s. "Wotan" served as Jung's endorsement of the German Faith Movement. Besides encouraging German Christians to join the Movement, he wrote:

> I would advise the German Faith Movement to thrown their prudery aside. Intelligent [more accurately, "Understanding"] people will not confuse them with those vulgar worshippers of Wotan whose faith is a mere pretence. There are people [representatives] of the German Faith Movement who are intelligent and human enough to believe and moreover to know that the god of the Germans is Wotan and not the universal Christian God. This is a tragic experience and no disgrace. (*Essays*, p. 15 [Hull deletes "universal"])

There have been a lot of superficial things written about Jung's "völkisch" sympathies but no one has bothered to follow closely Jung's argument in this, his most sustained völkisch text. R.F.C. Hull didn't help matters by blunting the "national/universal" dichotomy by deleting "universal" and changing "national" to "nationalist." When that dichotomy is restored we get to the crux of Jung's argument, which was that the various movements in Europe were expressing the *Volksseele* of the different countries, revolting against the universalist traditions of Christianity and the Enlightenment in favor of the countertraditions of aristocracy, ethnicity, and geographic particularity. It is the culmination of one strand in Jung's thinking that began with his university reading of Eduard von Hartmann and later of Arthur Drews and Leon Daudet. It created the common ground he shared with Hauer and Ninck and explains his differences with Keyserling whose cosmopolitanism finally seemed passé to him.

Although Hauer lost his position as leader of the Movement at the very moment the article appeared, he was not deterred from pursuing a career as a dedicated Nazi intellectual. In 1935 he had become a member of the League of University Lecturers. Along with Heyer he joined the Nazi Party in 1937 after it reopened membership and became a campus informant for the Security Service. Along with his colleague Max Wundt he would be temporarily barred from teaching after the war by the French authorities. He died in 1961.[47]

Hauer pursued his interest in race and religion by publishing books and conducting a series of "Aryan Seminars" at the university. This topic was the subject of an exchange of letters between Hauer and Jung. In his response dated June 7, 1937, Jung wrote:

> The connection between race and religion, which you have in mind, is a very difficult theme. Since the anthropological concept of race as an essentially biological factor remains completely unclarified, to demonstrate a connection between religion and this scarcely definable factor seems to me almost too bold an undertaking. I myself have personally treated very many Jews and know their psychology in its deepest recesses, so I can recognize the relation of their racial psychology to their religion...[48]

Here again Hull's translation subtly changes Jung's original words. A more accurate translation would be "so I can *surely* recognize a relation of their *special* religion to their racially-*conditioned* psychology..." (my italics). Hull deleted the words that Jung used to give emphasis to what sounds like a boast made to a Nazi intellectual. After Hauer's 1938 lecture "The Source of Belief and the Development of Religious Forms" to the Psychology Club the two had a falling out.[49]

Jung held an English-language seminar on Nietzsche's *Zarathustra* from 1934 to 1939.[50] Besides its detailed psychological analysis of the text, it was filled with Jung's numerous amplifications, digressions, and opinions about contemporary events. It is an invaluable source of additional information about topics Jung was discussing in his articles of the time. In the February 5, 1936, session, a month before "Wotan" appeared, Jung spoke about Hauer.

> I must say that I am very grateful to the Germans for their paganistic movement, at the head of which is my friend Professor Hauer who taught us the Tantric Yoga, and who is now become a savior of the fools. And some of them

> are so nice and honest; that they call it Wotan means of course that they are in a dream state where they cannot help telling the truth. (P. 813)

Jung's error in all this was his characterizing what was going on in Germany as natural and inevitable. In "Wotan" he compared the life of nations to boulders that crash down a hillside and are only stopped by an obstacle bigger than themselves. This naturalistic explanation and Jung's constant emphasis on the archetypal inevitability of events minimize the role that individual choice and collective activity have in shaping human affairs.

Another example from the seminar of Jung's penchant for ascribing "natural" motives to social conventions was that of racial mixture

> against which our instincts always set up a resistance. Sometimes one thinks it is snobbish prejudice, but it is an instinctive prejudice, and the fact is that if distant races are mixed, the fertility is very low, as one sees with the white and the negro; a negro woman very rarely conceives from a white man. If she does, a mulatto is the result and he is apt to be a bad character. The Malays are a very distant race, very remote from the white man, and the mixture of Malay and white is as a rule bad. (P. 643)

It should be noted that the comment was made on October 30, 1935, just six weeks after the Nazis promulgated the Nuremberg Laws that forbade marriage and all nonmarital relations between Jews and non-Jews. Although there is no direct reference to the Laws in the seminar, they were in the news and may have subliminally prompted Jung's remark. (Jolande Jacobi recalled that Jung had said to her "You know, I would never like to have children from a person who has Jewish blood."[51])

The editor's explanation that Jung presumably acquired these "genetic theories" during his medical school years is unsatisfactory. He locates the origins of Jung's views too far back in the past and credits them with a scientific rationale that they don't deserve. They were the product of a lifetime of personal feelings and intellectual influences but the specificity of Jung's examples indicate that his comments were derived from his old friend William McDougall. In his book *The Group Mind*, McDougall wrote about the crossing of races. "So the mulattoes...seem deficient in vitality and fertility, and the race does not maintain itself...Examples abound in Java of people of mixed Javanese and Dutch blood; and they are for the most part feeble specimens of humanity."[52] He went on to talk of inharmonious tendencies in the soul of crossbreeds in terms that Jung would use when he spoke about the multitude of ancestral units in the psyche that can dissociate. The book was published by Cambridge University Press in 1919 and may well have figured in the conversations the two men had when Jung visited London that year.

This interest in blood lines pops up in all kinds of places. Albert Oeri, Jung's oldest friend, was a newspaper editor and a politician deeply committed to the Swiss democratic tradition. In his contribution to Jung's *Festschrift*—"A Pair of Youthful Memories"—Oeri wrote about Jung's maternal family, the Preiswerks. After describing the scholarship of Jung's grandfather in the area of Hebrew philology he felt compelled to add the following disclaimer. "Otherwise the Preiswerks are a patrician family of Basel, and thoroughly Aryan." (More accurately, it reads "and of thoroughly Aryan descent."[53]) One wonders if Oeri, who did not attend Helly's séances with Jung because of his skeptical attitude toward spiritualism, was gently ridiculing his old friend's gullibility.

Press Coverage and Critics

Jung's work received periodic coverage in the German press. Heyer wrote an article for the *Kölnische Zeitung* (March 21, 1937) while his wife Lucy wrote one of several others that appeared in Cologne newspapers during the late 1930s. *Die Berlin Börsen-Zeitung,* which had published the Jennsen article in 1934, ran another one about Jung in 1936.

Some of the articles were reports of Eranos conferences and so focused on Jung's contributions to a psychological understanding of the Christian and Eastern spiritual traditions. Others dealt with Jung's unique approach to psychotherapy (von Hattingberg mentioned Jung's "theological blood"). Ernst Jahn, a Lutheran minister from Berlin, wrote "On the Weltanschauung Problem in C.G. Jung's Psychotherapy" for *Die Medizinische Welt* (July 20, 1935). In this lengthy article, Jahn discussed Jung's use of Eastern mystical thought to support his theory of personality; he also pointed out that this interest put Jung in the company of Eduard von Hartmann and Schopenhauer. Jung replied to Jahn in a letter dated September 7 saying that he was an empiricist and not a theologian. He went on to clarify that

> I do not by any means take my stand on Tao or any Yoga techniques, but I have found that Taoist philosophy as well as Yoga have very many parallels with the psychic processes we can observe in Western man...I have chiefly to do with people in whom I cannot implant any values or convictions from above downwards. Usually they are people whom I can only urge to go through their experiences and to organize them in a way that makes a tolerable existence possible.[54]

In 1936, *Rasse,* the monthly journal of the Nordic Movement, published a review of Martin Ninck's book on Wotan. The review began with a reference to Menzel's study of the god Wotan as the destiny of the Germanic race. The reviewer went on to say that Ninck, in the Romantic tradition of Bachofen and Burckhardt, "endeavors to press the old-Germanic *Weltanschauung* into the system of Ludwig Klages." This was a problem for the reviewer since in his opinion it emphasized the role of the Magna Mater at the expense of the warrior ideal.[55] Ninck contributed an article "Appearance and Expression" to the same journal later that year. It was a review of Klages' most recent book *The Fundamentals of the Science of Expression*. He began with a survey of the eighteenth-century conflict between Newtonian materialism and Kantian idealism. The impasse was resolved through the work of Goethe, Carus, and Nietzsche. It was, however, the work of Klages that established the proper relationship among body, soul, and spirit. Their relationships could only be understood through a study of their symbolic forms, a study that involved the fields of graphology, physiognomy, and depth psychology.

Jung's 1937 visit to Berlin was treated by the authorities as one of some significance. His presence on the reviewing stand at the parade and the press coverage he received attest to this. The reporter for the *Berliner Lokalanzeiger* (October 1) noted in her lengthy article that Jungian depth psychology came from a different "racial sphere" than Freud's psychoanalysis. His seminar at the Göring Institute was also reported on in the *Völkischer Beobachter*, the official newspaper of the Nazi Party on October 8. Entitled "The Archetypes," it noted that as the "original images" (*Urbilder*), they belonged to a long tradition in German philosophy. Symbols were important for a healthy functioning individual and for the interpretation of culture and the social life of a people.

In 1938 *Rasse* ran two articles on depth psychology that mentioned Jung. The first "Depth Psychology and the Nordic Race" (issue two) thanked him for freeing depth psychological understanding from the ghettoized narrowness of the Jews Freud and Adler. Besides his dream work, Jung's type theory would enrich racial science along the lines of Ferdinand Clauss' law of style of the Nordic race. The other "Depth Psychological Contributions to Racial Research" (issue ten) was similar in nature. "The most notable representative of depth psychology C.G. Jung" established the possibility of a reconstruction of cultural prehistory through his work on the unconscious mental processes. "Clauss and Jung have come, independent of each other, to many of the same results." He goes on to quote from Jung's 1935 Eranos address about the importance of religious symbols in the life of the soul.

In both articles Jung's name was linked to one of Germany's leading racial researchers—L.F. Clauss. Clauss was a prized contributor to the journal because of the books published by Lehmanns Verlag *Rasse und Seele* and *Die nordische Seele*. The first had a chapter "Racial Soul Science" in which the influence of Klages and Carus is evident with its preoccupation with the "gestalts of the soul." In his footnote to this, Clauss mentions Jung, Krueger, and Prinzhorn as well as Klages. In the second book he uses a photo of Klages as an example of a thinker of Lower Saxon descent. In a footnote Clauss refers to Jung's dichotomy of extraverted and introverted psychological types in support of his theory.

The same company that published *Rasse*, B.G. Teubner, also published the most popular book on human types to appear in Nazi Germany—H. Rohrbach's *Kleine Einführung in dir Charakterkunde* (*Short Introduction to Character Studies*).[56] He covered the work of Kretschmer on physique, and the more psychologically oriented systems of Jaensch, Jung, Klages, and Spranger. It is clear from the foregoing that Jung's work on symbolism and on types fit in nicely with the interests of a number of intellectuals in Nazi Germany.

Given their common philosophical heritage, Jung seems to have found what they said acceptable. One major objection he did have was to the biological emphasis of Nazi racial science. This is clear from the previously quoted letter to Hauer in which he reminded his correspondent that "the anthropological concept of race as an essentially biological factor remains completely unclarified." He continued this line of thinking when he declined a request from Egon Freiherr von Eickstedt, the editor of the *Zeitschrift für Rassenkunde*, for an article. "The connection between bodily disposition and psychic peculiarities is still so obscure to me that I cannot venture to speculate about it. My typology is concerned only with the basic forms of psychological attitude which I could not at present identify with any physiological or anatomical dispositions."[57] Jung was comfortable with the use of "race" as a category to the extent it conformed to the psycho-cultural definition he had held since his university years. He drew the line at the anthropological-biological definition promoted by the Nazis, which he found unacceptable because of its materialistic bias and misuse of genetics.

These developments in Jung's career were taking place against the backdrop of momentous events in the history of Europe. His article "Wotan" appeared in March 1936, the month that German troops entered the de-militarized Rhineland in violation of the Treaty of Versailles. His 1937 Berlin visit coincided with that of Mussolini, which marked another step on the road to the formation of the Rome-Berlin Axis.

Spain was the flashpoint for the ideological struggle that was shaping up across the continent. In 1931 the Second Spanish Republic was proclaimed but faced resistance from forces on both the left and the right. In 1932 Jung wrote "Who, for instance, would have dared to prophesy twenty years ago, or even ten, that Spain, the most Catholic of European countries, would undergo the tremendous mental revolution we are witnessing today? And yet it has broken out with the violence of a cataclysm."[58] In 1936 Franco began his military revolt and the Spanish Civil War was underway. To support his view of the situation there Jung quoted the Spanish philosopher Miguel de Unamuno (who immediately repudiated his remark),

> one of those Spanish liberals who undermined the traditional order in the hope of creating more freedom. Here is his most recent confession: "Times have changed. It is not any more a question of Liberalism and Democracy, Republic or Monarchy, Socialism or Capitalism. It is a question of civilization and barbarism. Civilization is now represented in Spain by General Franco's Army." Compulsory order seems preferable to the terror of chaos, at all events the lesser of two great evils. Orders, I am afraid, have to be heard in silence.[59]

Like other European conservatives Jung was willing to accept the authoritarian alternative to the socialist threat.

In this politically charged atmosphere Jung was scrutinized by commentators who reviewed his books that had recently been translated into Spanish. In 1935 Oliver Brachfeld's translation of *The Theory of Psychoanalysis* appeared and was reviewed in *El Sol* (Madrid, January 1, 1936). The anonymous reviewer noted that "Jung is a conservative psychologist in spite of the fact that his ideology is presented to us as liberal." Brachfeld joined the discussion with his review of *The Relation Between the Ego and the Unconscious* in the Catalan newspaper *Mirador* (Barcelona, April 16, 1936).

> The fact that among the great triumvirate of psychoanalysis, Freud, Adler, and Jung, only the latter can prove the "purity of his blood" and thus enjoy the sympathy of the dark Germany, favorably affects the promotion of Jung's works in the Germanic countries. In short, to be considered worth reading, C.G. Jung did not need the handicap of his two rivals. Regardless of his "racial equation," Jung is an author that writes well and has interesting ideas.

Perceptively, he pointed out the important footnote Jung added to the 1928 edition of the book in which he talked about it being an unforgivable error to consider the conclusions of a Jewish psychology as generally valid.

> He assumes a differentiation among the races, something that could only be explained in terms of climatic variations, environmental influences, etc. But if we admit these influences, why deny their effect on Jews, who, in today's Central Europe, cannot be considered anything but a very slight psychological (not racial) variety of white European humanity?

Two years later in the wake of the German annexation of Austria the *Vanguardia* of Barcelona ran the article "From Freud to Jung, or the Triumph of Zurich over Vienna" (April 2, 1938). "To fight Freudian doctrines, National Socialism does not need to oppose psychoanalysis. All it takes is to move away from the Jew Freud and his no less illustrious colleague Adler and take sides with Jung who seems to be an Aryan as pure as there can be." The reviewer seemed to be familiar with Jung's writings and activities and noted with a certain degree of irony that

> National Socialism, which claims to be a positive doctrine, could not accept such disturbing principles, and assailed them, attributing them to the "abject Semitic spirit." Jung was assigned to maintain the offensive within the field of psychoanalysis. But Jung's adaptation to the National Socialist doctrine has not been easy. The eminent Swiss psychologist was, of course, the most appropriate man to oppose the genial Viennese thinker. A man of Freud's caliber required a detractor endowed with exceptional qualities and Jung was. But in this case it was enough to be a pure Aryan.

By the late 1930s Jung's views and activities vis-à-vis Germany were well-known enough to be criticized by a number of other intellectuals. Among them were his former psychoanalytic colleagues who felt his current position only confirmed their long-standing suspicions about him. On October 15, 1937, commenting on Jung's invitation to deliver the Terry Lectures at Yale University, Rank wrote "Jung is coming next week to this country, seemingly as an apostle of Naziism. In today's issue of the *Saturday Review of Literature* he has an article on 'Wotan' justifying fascist ideology."[60]

Another psychoanalytic critic of Jung's was the German John Rittmeister.[61] First attracted to psychotherapy through the work of von Hattingberg, his studies took him to several European countries and finally to Switzerland. After working at the Burghölzli he joined the staff at a cantonal sanitorium in Münsingen. Suspected of communist sympathies he returned to Germany in 1937 where he became involved in the administration of the Göring Institute. Around that time he wrote a paper entitled "The Assumptions and Consequences of Jungian Archetypal Theory." He acknowledged the importance of Jung's emphasis on the role of the dialectical principle in the collective dimension of psychic activity. He was, however, critical of the Jungian tendency to withdraw from society and become preoccupied with subjective symbolic systems. The danger of this withdrawal was that the "god within" became the god of the bourgeoisie order of state and law. A preoccupation with an "organic, totalizing unity" becomes the front for a Volk community of slave-drivers and police. In the end, Jungian theory leads to solipsism and the deification of the ego, an elite, and a race. "In the holy crusade against corrosive science, against life-murdering Reason, C.G. Jung has appeared with a quick and elegant leap into the dark footlights of the political theater. There was no time to lose: people had long expected him in the struggle for the German World-soul."[62]

Rittmeister sought to ground his own theory and praxis in the humanistic tradition of Freudian psychology. By 1939 he and his new wife had become involved in a resistance group and were later arrested by the Gestapo. After spending time in prison, he was executed for treason in 1943. Rittmeister's effort to develop a theory that blended the ideas of Freud and Marx paralleled that made by a group of intellectuals affiliated with the Institute of Social Research, popularly known as the Frankfurt School.[63] The School developed Critical Theory, one of the dominant paradigms for analyzing the psychosocial dynamics of modern society and culture. The school adhered to the Enlightenment tenet of the primacy of reason in human thought and social organization. In light of Jung's often sarcastic dismissal of the power of human reason, it is no surprise that the School criticized Jung as one of the promoters of irrationalism who laid the intellectual groundwork for the triumph of National Socialism.

While on vacation in Italy in the summer of 1937 Walter Benjamin wrote to Gerhard Scholem that

> It is my desire to safeguard certain foundations of *Paris Arcades* methodologically, by waging an onslaught on the doctrines of Jung, especially those concerning archaic images and the collective unconscious. Apart from its internal methodological importance, this would have a more openly political one as well. Perhaps you have heard that Jung recently leaped to the rescue of the Aryan soul with a therapy reserved for it alone. My study of his essay volumes dating from the beginning of this decade—some of the individual essays date back to the preceding one—teaches me that these auxiliary services to Nationalist Socialism have been in the works for some time. I intend to make use of this occasion to analyze the peculiar figure of medical nihilism in literature: Benn, Céline, Jung.[64]

In a follow-up letter to Scholem a month later, Benjamin added "I have begun to delve into Jung's psychology... the devil's work through and through, which should be attacked with white magic."[65] Benjamin's critique was never written and his uncertain fate as an émigré came to its tragic conclusion in 1940 when he committed suicide at the Spanish border while fleeing the Nazi occupation of France.

It was Ernst Bloch who published the most sustained critique of Jung from the general perspective of the Frankfurt School in his *The Principle of Hope*. "But far more than with Bergson's 'élan vital,' the fascist Jung borders on the Romantic reactionary distortions which Bergson's vitalism underwent; as in sentimental penis-poets like D.H. Lawrence, in complete Tarzan philosophers like Ludwig Klages."[66] For Bloch, Jung and Klages oppose any progressive psychological development since both felt that the intellect undermined the instinctual basis of the imagination.[67] Like Rittmeister, Bloch acknowledged the importance of Jung's theory of the archetypes but faulted Jung for failing to extricate it from Romantic dilettantism.

> To fascism also, hatred of intelligence is, as Jung actually says "the only means of compensating for the damages of today's society." Fascism too needs the death-cult of a dolled-up primeval age to obstruct the future, to establish barbarism and to block revolution... the reactionary [Jung] wants to connect conscious material back with the repressed, to push it back ever deeper into the unconscious.[68]

In making his case Bloch misrepresented Jung's approach to the unconscious. In spite of some of his post-Freudian company and his biased remarks, Jung advocated a position that at its most humane and universal advocated a move *beyond* reason rather than a retreat *from* it. Jung hoped to help people cultivate a symbolic consciousness that was a product of a dynamic relationship between the conscious and the unconscious. The Frankfurt School took it as a given that Jung was a fascist intellectual, a charge that would become standard. In *Eros and Civilization* Herbert Marcuse relied on Edward Glover's *Freud or Jung?* (1950) to dismiss Jung as a reactionary.[69]

This view was also shared by the Surrealists who dismissed Jung for his deviation from Freudianism as well as for his right-wing politics.[70] Jung did, however, have one champion on the avant-garde scene during this period—Eugene Jolas (1894–1952), the founder of *transition*, the Parisian literary magazine that appeared from 1927 to 1938. In it he published Kandinsky, Duchamp, Joyce's "Works in Progress" (later *Finnegan's Wake*), and Jung who was also one of its active sponsors.[71] In a letter to James Oppenheim Jolas wrote,

> It seems to me that esthetic [*sic*] organization, or the "klare Bewusstheit" of the German romantics should be the final goal. The emergance [*sic*] of the phantasms with Jung is also merely a transitional-therapeutic-step towards full consciousness... What I have in mind is the development of a metaphysical-magical kind of literature in an age that is deliberately returning to the most facile naturalism and proletarian objectivism. I want to show the importance of Bachofen and his Mutter-Mythos, the breaking-up of language, the elements of the Gnostic in modern life, the entire complex of modern characterology.[72]

It seems that intellectuals were either intrigued or dismissive of Jung's interest in symbols.

In *Behemoth*, one of the first major studies of the Nazi state, Franz Neumann wrote "Even a National Socialist like the psychologist Jung (not to mention Nietzsche) is condemned for the dualism of his thinking."[73] Neumann was here responding to the thesis presented in *The Reich and the Sickness of*

European Culture (1938) by Christoph Steding. Steding was a fanatical young Nazi intellectual whose work was sponsored by the Reich Institute for the History of the New Germany founded by Walter Frank in 1935. A footnote in *Hitler's Professors* notes that

> This book, edited by Frank after the author's death, tended to show that for the Teutonic nations surrounding Germany the only way out of their spiritual crisis would be to unite under Germany's leadership into a Greater-Teutonic Reich. This book was so popular that in 1943 a third edition was issued.[74]

Steding excoriated intellectuals such as Johan Huizinga, author of *The Waning of the Middle Ages*, and Burckhardt from neutral countries such as Holland and Switzerland who had resisted appeals for the formation of a Greater German Reich. In particular he singled out Basel as having had the most pernicious influence on Germany's quest for a spiritually united Europe. He described with uncanny accuracy the milieu that he found objectionable. "From Jung's Zurich, from the Basel of Anthroposophy and the Egyptian enthusiasts of Bachofen... to this spiritual sphere belongs the Frankfurt of R. Wilhelm and Leo Frobenius which speaks of the East and Africa and even speaks of the primitive of the South Sea and South and North America."[75] He also criticized Keyserling for promoting, like Jews "of the Emil Ludwig type," a cosmopolitan spirit. What is ironic here is that although he was labeled a National Socialist by leftist intellectuals, it took a Nazi intellectual, one who could certainly recognize one of his own, to capture Jung's sometimes snobbish adherence to a conservative but essentially humanistic cultural agenda.[76] As we have seen this included a turn-of-the-century use of the word "race" that was distinct from the racial categories first promoted by German racial hygienists and later adopted by the Nazis. Although Jung generally used the term as a substitute for "nation" (as in the following "That is the way that Central Europe understood the psychology of the English race..."[77]) he did participate in the cultural-racialist discourse among German intellectuals that occurred during the Nazi period.

Jung's Reentry into the Anglo-American World

After 1935 Jung began to achieve a new level of public recognition in England and America. This stemmed from his status as one of the world's most famous psychiatrists and earned him honorary doctorates from Harvard, Yale, and Oxford for his lifetime of contributions to the field. On his visits he also gave public lectures and was sought out by journalists who solicited his observations on the increasingly volatile international situation. The companion piece to his interview with the American journalist H.R. Knickerbocker in the January 1939 issue of *Cosmopolitan* even featured Jung as the "Cosmopolite of the Month."

In the fall of 1935 Jung was invited to deliver a series of five lectures at London's Institute of Medical Psychology (the Tavistock Clinic) to an audience of about two hundred physicians. The founder of the clinic Dr. Hugh Crichton-Miller would have been favorably disposed toward Jung having heard about him from Maurice Nicoll with whom he started a practice in 1914. Besides presenting a survey of his approach to the human psyche and psychotherapy, the contacts Jung made there eventually led to the formation in 1938 of the English group of the International General Medical Society for Psychotherapy. Among the attendees were Jung's old associate and translator H.G. Baynes and a new acquaintance E.A. Bennet who later wrote several popular works on Jung's life and thought. Another participant was Wilfred Bion, later president of the British Psycho-Analytic Society who was accompanied by one of his patients, Samuel Beckett.[78]

Jung's engaging manner, enhanced by his command of colloquial English, proved to be such a hit that he was invited back the following year. The title of his lecture was "Psychology and National Problems" and dealt with his interpretation of current events, a topic that he had only touched upon the previous year. Jung focused on the traumatic impact World War I had on Russia, Germany, and Italy. The misery and distress that they experienced led to an emotional regression on a collective level. This process did not stop at infantile modes of behavior but went back to archaic patterns of thinking that, in the case of Germany, coalesced around the charismatic figure of Adolf Hitler who functioned as a medicine man promising salvation through allegiance to his mystical doctrine.

Much of Jung's analysis focused on the dominant role that the state had come to play in the life of modern society. "The State is the psychological mirror-image of the democracy monster... it squeezes its contributions out of the most vital and gifted individuals of its domain, making slaves of them for its own wasteful purposes."[79] A little later he mentioned how taxation had made the great estates of England uninhabitable. Like a good Swiss he was infuriated by the monetary policies being followed, especially the decision to go off the international gold standard. Money was being hollowed out, which would make savings illusory and would rupture cultural continuity and creativity that depended on responsible and independent individuals.

Jung included his observations about New Deal America, having just returned from the Harvard Tercentenary where he heard Franklin Roosevelt deliver the keynote address. His comments confirm his essentially conservative outlook, and sound like what Roosevelt's Republican critics were saying.

> But if you carefully study what President Roosevelt is up to and what the famous N.R.A. [National Recovery Act] meant to the world of American commerce

> and industry, then you get a certain idea of how near the great State in America is to becoming Roosevelt's incarnation. Roosevelt is the stuff all right...[80]

Just what that stuff *was* is revealed in an interview that Jung gave to the *Observer* (London).

> I have just come from America, where I saw Roosevelt. Make no mistake, he is a force—a man of superior and impenetrable mind, but perfectly ruthless, a highly versatile mind which you cannot foresee. He has the most amazing power complex, the Mussolini substance, the stuff of a dictator absolutely.[81]

At the same time, Jung noted the widening gulf between left and right, with the choice being between chaos and enforced order. Applying his homeostatic model of psychic functioning to social dynamics, he said,

> Communistic or Socialistic democracy is an upheaval of the unfit against attempts at order...In as much as the European nations are incapable of living in a chronic state of disorder, they will make attempts at enforced order, or Fascism...After the dictators? Oligarchy in some form. A decent oligarchy—call it an aristocracy if you like—is the most ideal form of government. It depends on the quality of a nation whether they evolve a decent oligarchy or not. I am not sure that Russia will, but Germany and Italy have a chance.

Jung was compromising his Basel allegiance to a spiritual aristocratic principle by still looking with hope on the criminal elites that controlled Germany and Italy. In this he was closer in outlook to his new sponsor Karl Anton Rohan than he was to his old friend Albert Oeri.

Based on his familiarity with Jung's ideas H.G. Baynes wrote a book called *Germany Possessed* (1941). Basically it is an elaboration of Jung's interpretation of Germany's psychological development and Hitler's shamanic role in it. What is interesting is that he relied heavily on Herman Rauschning's *Hitler Speaks* (1940) for anecdotes about the dictator. Rauschning's involvement in the Nazi Party had led to his becoming president of the Danzig Senate. He later left the party, emigrated, and became one of most important conservative critics of Nazi Germany. Rauschning wrote the introduction to *Germany Possessed* and in it expressed a viewpoint strongly influenced by Jung ("The question arises whether Hitler is not himself the expression of the shadow-side of our whole civilization"[82]). Rauschning had been familiar with Jung's work for some time and here his debt is most obvious. His shared political philosophy is evident when he writes "For the author shows how the danger of self-destruction in Germany does not arise merely from the revolutionary dynamism of National Socialism, but also from the lack of effective opposition of conservative forces."[83] As we shall see Jung was to have significant contacts with that conservative opposition whose efforts culminated in the 1944 bomb plot.

Jung's efforts to truly internationalize the International General Medical Society for Psychotherapy first bore fruit in 1937 when the first conference to be held outside Germany took place in Copenhagen. It took another major step forward with the founding of the English group at the Oxford Conference in 1938. With one hundred fifty members, it immediately became the second largest national group after the Germans. Although the conference themes of psychosomatic medicine and the stages of life were noncontroversial, the German delegation was impacted by official government policies: Göring complained to Jung about the fact that Strauss, the president of the new group, was Jewish; also several members, Heyer among them, were not given permission to attend. Heyer did, however, attend that year's Eranos conference on the Great Mother, the first time he had done so since 1935.

In 1936 Jung traveled with his wife Emma to the United States to attend the tercentenary celebration of Harvard University. He delivered a paper "Psychological Factors Determining Human Behavior" (CW 8) and in the company of sixty-five other scholars and scientists from eighteen nations received an honorary degree. Apparently Freud was the psychology department's first choice but concerns that he would decline due to his poor health prompted them to opt for Jung.

The faculty member who lobbied most actively for Jung was Henry A. Murray (1893–1988) who had received a degree from Harvard in 1915. After he was married he met Christiana Morgan who sparked his interest in Jung and with whom he began a lifelong affair.[84] Murray went to Zurich in 1925 and was transformed by his encounter with Jung. He later cofounded Harvard's Psychological Clinic whose research projects resulted in *Explorations in Personality* (1938), a landmark in American psychology. Morgan is an important figure in her own right. With Murray she created the Thematic Apperception Test, which grew out of her own experiences of active imagination, the subject of Jung's Vision Seminars (1930–1934).[85]

Murray came to Jung's defense in the pages of the student newspaper *Harvard Crimson* when questions about Jung's compromising with the Nazis were raised. In the May 27, 1936, issue a quote from Jung's first *Zentralblatt* editorial about the differences between Germanic and Jewish psychology was juxtaposed with Göring's endorsement of *Mein Kampf* as the basis for the German Society's scientific work. Two days later Murray responded by pointing out that the juxtaposition was misleading and that it was important to quote Jung's concluding remark in which he stated that no inferiority of the Semitic psychology was implied.

> Dr. Jung is a thorough-going Swiss—bluff, independent, wise and utterly aloof from political entanglements. To a mind of such universality the Nazi racket is a phenomenon to be impersonally studied, and perhaps judged from an emotional

> distance. That he should be persuaded to pay lip service to the present German regime cannot be supposed by anyone who knows him.

By 1936 the view that Jung was a Nazi sympathizer was widespread; he was sensitive enough to the charge that he had prepared a press release for his visit in which he stated that he detested politics and was neither a Bolshevik, a Nazi, nor an anti-Semite (CW 18, pp. 564–565). In a letter written shortly after the tercentenary to Abraham Aaron Roback, Jung repeated that he was not a Nazi and explained his reason for accepting the presidency of the International General Medical Society for Psychotherapy. He did it in order to protect psychotherapy in Germany and reorganized it in order to help Jewish members maintain at least some professional affiliation. He did acknowledge that he had insisted on recognizing the difference between Jewish and Christian psychology since 1917. This was evasive because in his article "The Role of the Unconscious" Jung had contrasted Jewish psychology with Aryan not Christian psychology and so glossed over the fact that he was basing the difference on racial not religious factors. (This article was one of the few that had not been translated into English prior to its appearance in the Collected Works in 1964, three years after Jung's death.)

Roback soon afterward sent Jung a copy of his book *Jewish Influences in Modern Thought* (1929) who responded with a letter in which he expressed the opinions he had held for many years ("I'm quite aware of the fact that Freud's statement is necessary for the Jew..."[86]). Roback (1890–1965) was a psychologist who was a student of William McDougall's at Harvard and published his massive *Psychology of Character* in 1927. It drew not only on Jung's *Psychological Types* but also on Beatrice Hinkle's modification of that typology that Roback noted had been adopted by McDougall in his textbook on abnormal psychology. Along with Jung Roback was a contributor to the first issue of McDougall's journal *Character and Personality* in 1932. He is the likely source of the note to it about Murray's Harvard Psychological Clinic analyzing dreams about the kidnapping of the Lindbergh baby.

It was Henry Murray who was the locus of the conflicting criticisms and defenses of Jung. He was pressed about Jung's Nazi connections by Felix Frankfurter, then on the faculty of Harvard Law School, but soon to be appointed to the U.S. Supreme Court by Franklin Roosevelt.[87] At the same time, Murray received several letters from Jung that reveal his extreme sensitivity to the allegations being made against him. Jung had heard from Hinkle that Dr. Hadley Cantril of Princeton University had been spreading the rumor that he was a frequent visitor to Hitler's mountain retreat at Brechtesgaden. Jung also mentioned difficulties that he had encountered at Oxford and on his trip to India that year (1938). The latter incident reveals the very streak of paranoia that Jung explicitly denied was at issue. He said that while there he was shown a faked photograph sent years before from Vienna. It depicted him as "a Jew of the particularly vicious kind."[88] Why would Freudians want to caricature Jung as a "vicious Jew" and then send it off to India of all places? The incident makes clear the degree to which Jung felt that he was still being punished by a psychoanalytic cabal for his break with Freud.

In 1937 Jung was invited to deliver the Terry Lectures at Yale University. In his paper "Psychology and Religion" he delineated his understanding of the natural religious function of the psyche as manifested in both an individual's dreams and in the collective life of a nation. For the latter he drew on the analysis of Germany he made in his "Wotan" article, a version of which was in the *Saturday Review of Literature*. He focused on the case of Nietzsche (the subject of his ongoing English-language seminar) and characterized his life as an expression of the latent Wotan archetype that was gaining ascendancy in Germany, especially as a result of the World War. One comment reveals again the basic flaw in Jung's analysis. "Those Germans were by no means people who had studied *Thus Spake Zarathustra*, and certainly the young people who resurrected the pagan sacrifices of sheep knew nothing of Nietzsche's experience."[89] This is not the case, however. Nietzsche had become a cultural icon and major influence on German intellectual life after his death in 1900. His philosophy infused the German Youth Movement and as the old story goes, every German soldier marched off in 1914 with a copy of *Zarathustra* in his knapsack. With his preoccupation with establishing the archetypal origin of the Wotan experience he was unable to see the extent to which this phenomenon had been consciously incorporated into a völkisch ideology. Interestingly, the two authors Jung cited in the footnote to this quote prove this point: Baeumler became the leading Nietzsche expert in Nazi Germany and Elisabeth Förster-Nietzsche controlled her brother's estate, which she used to promote her anti-Semitic, nationalistic point of view.

The January 1939 issue of *Cosmopolitan* featured Jung's interview with H.R. Knickerbocker, one of the leading international correspondents of the time. It was a long and detailed psychological analysis of the Hitler, Mussolini, and Stalin. Jung characterized Hitler as the medicine man leader in contrast to the other two who fit the profile of the chieftain type. This was because Hitler was highly susceptible to information coming from the unconscious and so was in tune with the collective unconscious of the German people. His inner voice guided him to success in face of the doubts of his advisors. Jung's remarks show just how closely he was monitoring such political developments in Europe as German rearmament and the Czech crisis.

At one point Knickerbocker asked Jung why foreigners seemed immune to Hitler's charisma. Jung responded "It is because Hitler is the mirror of every German's unconscious, but of course he mirrors nothing from a non-German."[90] Although Jung was speaking more critically of Hitler, at least to an English-speaking audience, he was deeply involved on an unconscious level as we shall soon see in a dream that Jung had around this time. Since Jung had talked about his *participation mystique* with what was going on in Nazi Germany during his Tavistock Lectures this comment reminds us the extent to which Jung did identify with his Germanic roots.[91]

Toward the end of the interview Jung the physician shifted from diagnosis to his recommendation for treatment. He suggested that the best thing that the Western powers could do was to encourage Hitler to attack the Soviet Union.

> I say let him go to the East. Turn his attention away from the West, or rather, encourage him to keep it turned away. Let him go to Russia. That is the logical *cure* for Hitler... There is plenty of land there—one sixth of the surface of the earth. It wouldn't matter to Russia if someone took a bite, and as I said, nobody has ever prospered who did.[92]

This position, which he was to reiterate in a wartime interview, presented his conservative, anticommunist stance in clear relief and nicely dovetailed with the Nazi goal of *Lebensraum* ("living space"), first proposed in the 1920s by Munich professor Karl Haushofer. This was not the first time that Jung supported an aggressive German foreign policy measure.

> When they broke into Belgium [in 1914] they said yes, we have violated the Treaty; yes, it *is* mean. That is what Bethmann-Hollweg always said: "We *have* broken our word," he confessed. And then we said how cynical he was, and that the Germans were only pagans anyway. But they simply admit what the others think and do.[93]

Remember that Jung actually witnessed the German invasion on his way back to Switzerland from Britain and was deeply impressed by the social solidarity he saw and termed a "feast of love." It made a lasting impression and was a personal experience of the Wotanic surge that he saw developing in Germany.[94] The Wotanic dynamics apparent in Nazi Germany were the result of a carefully orchestrated script directed by Goebbels and his Ministry of Propaganda. While German industry and the Wehrmacht were preparing for war Hitler engineered a series of diplomatic coups that heightened anxieties about another European war: the remilitarization of the Rhineland (1936), the annexation of Austria (1938), and the Czech crisis (1938).[95] Jung favored Chamberlain's policy of appeasement since he felt it unwise to confront Nazi Germany directly.

On August 23, 1939, the world was stunned by the news that Hitler and Stalin had signed a nonaggression pact. This deal sealed the fate of Poland, which was invaded nine days later. This was personally troubling to Jung since he had publicly advocated a German invasion of Russia (like other conservatives of the time he refused to call it the Soviet Union and in his 1934 "Rejoinder to Bally" still referred to Leningrad as Petersburg). Shortly after the announcement Jung had an important dream that he reported to Esther Harding in 1948 and in modified form to E.A. Bennet two years earlier.[96]

> He found himself in a castle, all the walls and buildings of which were made of trinitrotoluene (dynamite). Hitler came in and was treated as divine. Hitler stood on a mound as for a review. C.G. was placed on a corresponding mound. then the parade ground began to fill with buffalo or yak steers, which crowded into the enclosed space from one end. The herd was filled with nervous tension and moved about restlessly. Then he saw that one cow was alone, apparently sick. Hitler was concerned about this cow and asked C.G. what he thought of it. C.G. said, "It is obviously very sick." At this point, Cossacks rode in at the back and began to drive the herd off. He awoke and felt "It is all right."

Jung's associations were as follows: Hitler was the Anti-Christ, the herd represented the disturbed instincts expressed in the one-sided masculine ideology of the Nazis, while the Cossacks represented the sounder instincts that would overthrow Germany. In the Bennet version the dynamite castle becomes some barracks in a field, there is no mention of the sick cow, and instead of a consultation, Jung felt that all would be well as long as he fixed his gaze on Hitler.

The first thing that is significant is how Jung treated two of the dream's most important elements—the dynamite castle and the sick cow. With Harding he discussed them in relation to the German collective situation. The fact that he deleted them in the version he gave to Bennet indicates some sensitivity to their subjective meaning. His strategy was to convert an anxiety dream (Cossack revenge for his advocacy of an invasion of Russia) into another example of his psychic sensitivity regarding current events.

Many people are familiar with his pre–World War I visions of a European bloodbath but aren't aware of the fact that in the first years of that conflict he was constantly dreaming of interviewing the kaiser.[97] Jung felt a special connection to developments in Germany and imagined himself to be a German shaman, a role that would put him in the company of Nietzsche and Hitler. This comes across vividly in the image of Hitler and Jung being raised up on equivalent mounds (power spots). Like two old medicine men they consult each other over a sick cow. This could be connected to Jung's experience of being on the reviewing stand in Berlin in 1937. Jung tried to salvage some of his heroic self-image in his account to Bennet by substituting an apotropaic gaze as a magical defense against Hitler in place of his collaboration/consultation with the dictator.

The mound symbolically expresses Jung's sense of Nietzschean elitism that elevates the few above the herd. The image of the herd also figured in a dream that Jung reported in his 1925 seminar. "In 1910 I had a dream of a Gothic cathedral in which Mass was being celebrated. Suddenly the whole side wall of the cathedral caved in, and herds of cattle, with ringing bells, trooped into the church."[98] This came after a long amplification of his Elijah/Salome fantasies and a discussion of the Mithraic mysteries. In both dreams, a medieval (outmoded) structure is suddenly filled with a herd of cattle associated with masculine energy (the Mithraic mysteries were open only to men and were popular among Roman soldiers). By the late 1930s, Jung had come to see the Hitler religion as analogous to the rise of Islam, a militant, expansionary movement.

This can be amplified by a passage from the Zarathustra Seminar that followed a discussion of the relationship of the inferior function to the mob. Our natural inclination is to avoid the inferior function but the experience is necessary for psychological growth. The alternative is "Nietzsche's aristocratic attitude [that] has the tendency to travel to Mt. Everest and to get frozen to death..." The church can for a time act as a necessary container for the herd but "At other times, the prison or the stable is no longer satisfactory. For instance, if the herd has grown and there are too much [*sic*] head of cattle, then the moral demands must be lowered, because the greater the crowd, the more immoral and archaic it is..."[99] After further discussion, Jung said

> [People] have dreams of high tension wires that should not be touched, or dynamite or a strong poison or dangerous animals or a volcano that might explode. Then one has to warn people and take them a safe distance away from the place where they touch the high tension which would overwhelm them.[100]

The sick cow that is isolated in the dream and ignored by Jung in his interpretation is the key to understanding his potentially explosive psychic situation. From a Jungian perspective the cow would represent the neglected feminine, neglected not just by the Nazis but by Jung whose conscious advocacy of the *Lebensraum* policy reveals a callous disregard for the immense human suffering that would soon be its consequence. Although in this dream Jung was still identifying with Hitler's grandiosity he had begun to take a more critical view of developments in Germany. What were the reasons for this? Feedback he would have been getting from American and British friends and colleagues in the late 1930s would have begun to make him reconsider his opinions. Furthermore, his 1937 visit to Berlin gave him a chance to see Hitler up close and observe the regime's preparations for war. Contacts with two Germans outside his established circles also made him reconsider his earlier positive opinion of the Nazis. James Kirsch identified that year as the turning point of Jung's thinking. Still, we shall see that Jung did not fully begin to process the unconscious material found in the dream until he was recovering from his 1944 heart attack and then writing *Answer to Job* after the war.

Chapter 6

The World War II Years

For the second time in the century Europe was caught up in a devastating war and for the second time Switzerland managed to avoid being a participant. Eventually the country was completely surrounded by Axis-controlled territory and had to devise a policy of survival. It seems clear in the wake of the "Swiss gold" controversy that the postwar myth of Swiss neutrality during World War II cannot be sustained. The seven-member Federal Council's willingness to cooperate with the Nazis owed more to the pro-German sympathies of some of its members and to the profits to be reaped from cooperating with the Nazis than to any real threat to the country. The Nazis never seriously considered launching an invasion; Switzerland was far more valuable as a financial clearing house and arms supplier than as an occupied territory.[1] Nor were the Swiss particularly welcoming to Jewish refugees; it was Swiss authorities who, fearing the "Judaization" of Switzerland, lobbied Berlin to have the passports of Jews stamped with the letter "J" to help distinguish Jewish refugees in the wake of the annexation of Austria. Only about twenty thousand Jews were admitted after November 1938. They were interred in camps and had their expenses paid by special assessments levied on the resident Jewish community. By 1942 the Swiss border was effectively closed to Jewish refugees who were forced back into the waiting arms of the Gestapo. Even the group around Jung was affected by this fear. In December 1944 anxieties about a "Jewish influx" led the executive committee of the Psychology Club of Zurich to codify a long-standing unwritten quota that limited Jewish membership to no more than 10 percent.[2]

Although the Swiss government capitalized on the conservative, insular attitudes of the majority of the population, its overtly pro-Axis policies did not reflect popular sentiment and were, in fact, not made public to the Swiss people. This was not only due to press censorship but also to the unique structure of the Swiss government. It consists of a two-house legislature: the National Council whose members are elected every four years and the Cantonal Council that represents the interests of the cantons. They in turn elect the seven-member Federal Council that functions as both a cabinet and a collective executive with each minister assuming the largely ceremonial presidency for a year on a rotating basis.

During the war the Federal Council was given extraordinary powers that were not subject to parliamentary review; each minister held almost proprietary power over his ministry. The most notorious exemplar of this was Marcel Pilet-Golaz, the minister of foreign affairs (1940–1944) and a Vichy sympathizer. His speech on June 25, 1940, immediately after the fall of France discussed Switzerland's place in the Europe of Hitler's New Order. Though public criticism followed swiftly, Pilet-Golaz did not experience a change of heart but merely changed his tactics. He proceeded to secretly negotiate an economic accord that effectively coordinated Switzerland's economy with the Third Reich's. In exchange for coal, iron, and foodstuffs Switzerland supplied weapons, ball bearings, and optics as well as critical financial services and guaranteed usage of the St. Gotthard Tunnel that carried supplies to German forces in Italy.

What is uniquely Swiss about the Federal Council was the extent to which it reflected the country's social conformity and adherence to a consensus that is forged at the price of critical debate. Switzerland is a small, landlocked country comprised of four language/cultural groups, two major religious denominations, and a long history of nonparticipation in the political affairs of Europe. Its citizen army has promoted a sense of national identity while its officer corps functioned as a clubby network of men who are the country's bankers, businessmen, and politicians. Although the commander-in-chief Henri Guisan and the average Swiss soldier were solidly pro-Allied, many in the officer corps openly admired the German military even in its Nazi incarnation. Eugen Bircher, a high-ranking medical officer, sponsored a series of medical teams that served with the German army on the Eastern Front (remember that he had also served with the Germans in World War I). In 1918 he had founded the Swiss Fatherland Association, a superpatriotic organization that now closely monitored the "Jewish Question" and successfully lobbied the Federal Council to adopt its highly restrictive refugee policy.[3]

The extensive powers retained by each canton and enumerated in the constitution created a strong interest in local affairs that inhibited the formation of political parties with a truly national vision and appeal. Most parties with a national voice were alliances of cantonal parties that represented various special interest groups (e.g., the Farmers Party and the Catholic Conservative Party). The Social Democratic Party represented, of course, the interests of the working class. The Liberal-Democratic Party was strongly supported in French-speaking Switzerland and in Basel where it was headed by Jung's old friend Albert Oeri who was the editor of the *Basler Nachtrichten* and a member of the National Council and a fierce defender of the freedom of the press. This all provides the historical context for Jung's activities during the war. Forced to curtail his many international activities, he retreated into a world of scholarship and spent more time than ever at his tower.

Jung's Political Activities and Views

All too many students of Jung have taken his self-designation as being "unpolitical" at face value—but the truth is that Jung was sufficiently involved in politics to run for a seat in the National Council in the fall of 1939 (he lost). He was a candidate for the Landesring der Unabhandigegen (National Group of Independents), a party started by Gottlieb Duttweiler.[4] Duttweiler's path to success was an unusual one for Switzerland. After spending time in South America as a young man he returned home and began to sell produce from the back of a truck. His success led to the creation of Migros, a farmers' cooperative that built a chain of supermarkets found throughout the country today. A democratic populist, he founded the Landesring to promote the interests of the "little man" who had been ignored by the traditional parties, including the liberal Freisinnige Partei. Jung suggested to Duttweiller that wage adjustments be enacted in order to relieve the financial burden being carried by the thousands of men who had been called to active duty (450,000 at its peak).

Jung apparently maintained some contacts with the Zurich wing of the Freisinnige Partei since in 1943 he was invited to give a lecture to the party's group in Kusnacht, a wealthy Zurich suburb where he lived for his entire adult life.[5] The lecture was entitled "Observations and Thoughts about the Present Cultural Crisis" and reflected something Jung had written to describe his political candidacy: "I'm told people want representatives who represent spiritual values."[6] In Jung's opinion, the crisis had its modern origin in the separation of religion and science. At fault was the materialistic *Weltanschauung* exemplified by Ludwig Büchner that had created a preoccupation with material well-being not only in the proletariat but also in the general population. This cultural crisis has led to a serious illness in the body of the nation. "We Swiss must open our eyes and not be pharisaical. We must ask ourselves: how could we oppose the power-principle if we were a small state of one hundred million?" When divine authority is challenged the "lord of this world" as described by John the Evangelist appears, or to put it more succinctly, "the rule of the devil." The lecture ended on a rather cryptic note.

> The doctor, having to withhold the truth from his patient, should often confess "I must lie." Guilt remains guilt, even for a noble purpose. As in the life of the

individual there is a higher justice in the life of a people. If there is no more value, then the cultural crisis is thrown wide open.

Although it is difficult to sort out this mix of theological metaphors and medical diagnosis he seems to be saying that the materialistic power-drive is "the devil" and threatens to subvert the cultural ideals that sustain a society.

These sentiments were also expressed in an interview Jung gave the year before to the *Schweizer Illustrierte Zeitung*.[7] Again Jung assumed the persona of a consulting doctor to discuss the "European fever" that had beset the continent. He made no reference to Germany's culpability in starting the war but again chose to trace the cause of the malaise to nineteenth-century materialism and its deleterious effects on cultural and spiritual traditions. In particular he singled out Sigmund Freud as "the mouthpiece of the uprising of sexuality." Because of the "unrestricted and unrestrained living out of the urges, we were exposed to a catastrophe of unforeseeable consequences." Prompted by Switzerland's wartime isolation and the fact that his words were reaching his fellow Swiss via the popular press, Jung's remarks convey a decidedly patriotic-pious message. It is "our holiest obligation to resolutely hold on to the handed-down values if all the products of art, science, and morals are not to be washed away by floods of passion and waves of thoughtlessness." Jung was expressing a more sober Burckhardtian message quite different in tone than his previous Nietzschean pronouncements about the German phenomenon. He closed the interview with the image of Switzerland as an island of contemplation amidst the chaotic sludge, a monastic refuge from the modern Dark Age.

Jung did not acknowledge the extent to which his professed medical and political neutrality masked a conservative political agenda. He glibly applied the "laws of nature" to historical developments with the result that war is described as an inevitable product of the self-regulating system of international relations. When the reporter asked him if the war could have been avoided he said "certainly not, but certainly the catastrophe which has descended on Western Europe." Reiterating what he said in the 1939 Knickerbocker interview, he said that it had been a mistake for the Western powers to try hindering Germany's expansion to the east. This was expressed three years after the German blitzkrieg had devastated Poland and brought unimaginable suffering to Eastern Europe. Like all European conservatives he supported a German crusade against atheistic communism. His conservative views and respect for Great Power politics is evident in his observation about the reason for Great Britain's long-term success. Rather than emphasizing the country's multiparty system and tradition of dissent, he lauded the country's enlightened imperialist policies, "which always skillfully understood how not to let the subordinated peoples feel their dependency." Jung's "nonpolitical" persona should not distract us from the fact that he respected, even admired those who exercised political power. His *Realpolitik* attitude was a likely psychogenic factor in the heart attack and hospital visions he experienced in 1944.[8]

The Wartime Eranos Conferences and Jung's Paracelsan Interests

The onset of the war disrupted the network of students and colleagues that had gravitated to Zurich to be near Jung. Americans such as Paul and Mary Mellon returned home when the invasion of France became imminent and were followed by German émigrés such as Heinrich Zimmer who eventually reached New York after a stay-over in England.[9] The war also impacted the line-up of presenters at the annual Eranos conference. Most prewar regulars were no longer available, so Olga Froebe-Kapteyn turned to a group of mostly Swiss scholars to help carry on the conference. They included Jung's old colleague J.B. Lang, Max Pulver, a graphologist and Psychology Club Zurich member, and Walter Wili, a professor of classics at the university in Bern. They were joined by another classicist, the Hungarian Karl Kerenyi, who was serving as a cultural attaché in Switzerland at the time and who would become an Eranos stalwart.

Kerenyi's appearance signaled other interesting links to Jung. McGuire noted that "since 1929 he had been a pupil of the noted classicist W.F. Otto"[10]; Otto's two most famous books available in English are *Gods of Greece* (1929) and *Dionysos: Myth and Cult* (1933). The latter was first published as the fourth volume in the series "Frankfurter Studies in the Religion and Culture of Antiquity." Otto, who spoke about the Eleusinian Mysteries at the 1939 Eranos conference, was on the faculty of the Frankfurt University and a member of the group that had gathered around the philosopher Kurt Riezler who was also the university's rector. A former aide to Bethmann-Hollweg, Riezler later emigrated to the United States where he joined the New School for Social Research, Graduate Faculty in New York. Another member of the group Ernst Kantorowicz had, like Otto, previous contacts with the Stefan George Circle. Finally, Otto knew Richard Wilhelm whose China Institute was affiliated with the university. In general this group shared Jung's culturally conservative humanistic philosophy.

In honor of Otto's sixtieth birthday Kerenyi gave a lecture at Frobenius' Research Institute for Cultural Morphology on June 22, 1934. This venue was a most appropriate one since both Otto and Frobenius were members of the Doorn Research Community. This was a small group of scholars who had gathered around ex-Kaiser Wilhelm II who lived in modest exile in Doorn, Holland, where they met for an annual conference. Their research topics and methodology were derived from Frobenius.[11] The Kaiser himself eventually gave lectures that were published as *The Chinese Monad* (1934) and *Studies of the Gorgo* (1936), both of which traced the diffusion of symbols from cultural zone to cultural zone with special attention to the swastika motif.[12] Jung owned a copy of the first book, apparently given to him by a follower named Erika Schlegel. Jung began to meet members of this group through Count Keyserling and the School of Wisdom. He and Frobenius spoke at the 1927 Conference and his travels to Africa and the American Southwest were partly inspired by the many expeditions Frobenius made around the world.[13] Besides meeting Frobenius at Keyserling's, Jung also met the ex-kaiser's brother Heinrich who regaled him with his account of escaping the Sparticist uprising in Berlin after the war.[14]

Another point of intersection for Jung, Otto, and Kerenyi was Prince Karl Anton's *Europäische Revue* in which they all had got published. Rohan had easily accommodated his neoconservative views and fascist sympathies to Nazi ideology and his journal continued into the war years until it finally succumbed to the paper shortage brought on by Germany's total war economy. Jung's affiliation with the *Revue* had ended in 1934 when his work began to appear regularly in the *Zentralblatt* and in the Eranos Yearbooks. Otto's "The Humanity of the Greeks and Posterity" appeared in the August 1937 issue and included references to Ninck's book on Wotan. The *Revue* published four articles by Kerenyi: "On the Crisis and Possibility of a Science of Classical Antiquity" (December 1937), "What is Mythology?" (June 1939), "Platonism" (October 1941), and "The Secret of the High Cities" (July/August 1942).

This last article dealt with the sacral space of ancient Roman cities and referred to *Italy and Rome*, the recently published book of another of Otto's students in Frankfurt, Franz Altheim, who had worked for Himmler's *Ahnenerbe* (Ancestral Heritage Institute).[15] The book was published in 1941 by Pantheon Akademische Verlaganstalt of Amsterdam and Leipzig. This is of particular interest because at the same time the company released two collaborations by Jung and Kerenyi, *The Divine Child* and *The Divine Maiden*. They were published by the Amsterdam branch of the company, which seems to indicate the ongoing influence of Otto through his Doorn Research Community connections. The fact that they were not published in Leipzig stems from the fate of Jung's works in Germany during the war.[16] In 1942 and 1943, L. Fernau, the book distributor for the Leipzig branch of Jung's regular publisher Rascher, wrote letters discussing the unavailability of Jung's books in Germany due to censorship.[17] In occupied Paris Jung's *Essays on Analytical Psychology* appeared on a list of banned books that was issued in 1940 and updated in 1942.[18]

The picture of Eranos and related publications during World War II that emerges is an ambiguous one. Any advantages German cultural authorities might have gained from Jung's reputation or writings were rendered marginal by the rapid development of events after 1939. For his part, we shall see that after the war Jung used the fact that his books had been banned and his name put on a Gestapo list of possible Swiss detainees (as yet unsubstantiated) to portray himself as an anti-Nazi. It should be noted that to be targeted by the Nazis did not make one ipso facto an *anti*-Nazi. All sorts of people fell victim to them for reasons that often had little to do with political opposition; many were shocked to find that they had run afoul of some party agency that was often pursuing an agenda at odds with those of other bureaucracies. (One famous example is the Expressionist artist Emil Nolde. In spite of being an early member of the Nazi Party he found his work lumped with other "degenerate" artists and was forced into seclusion.)

Another wartime Swiss participant of Eranos was the Bern classicist Walter Wili who lectured in 1943–1944 through 1945 and contributed a piece to the *Festgabe* for Jung's seventieth birthday in 1945. He followed Kerenyi in the speaking order each year and, while not a student of Otto's, was affiliated with his circle. This is clear from his article "A View on Antiquity" that appeared in the October and November 1941 issues of the *Europäische Revue*. In it, he cited Otto and the Albae Vigiliae series published by Pantheon that included works by Altheim as well as the Jung-Kerenyi *Divine Child*. The classical scholarship of these conservative humanists was judged acceptable by Nazi censors who could find nothing suspicious in the apolitical content of ancient Greek and Roman religion.

A closer look at Wili's career does make it clear, however, that his conservatism was not just that of an apolitical classicist. He had received his PhD from the University of Zurich in 1930 where, in light of his later activities, it is possible that he was involved with the student group that gathered around Robert Tobler and formed the National Front, Switzerland's first major fascist organization. He was appointed to the faculty in Bern in 1933 and was a founder of the *Bund für Volk and Heimat* ("The League for the People and the Homeland"), which was less stridently anti-Semitic than the Front but stressed the superiority of Christian culture.[19] He expressed his concerns in his 1934 book *Switzerland and the Fate of Europe* and summarized them in "Spiritual National Defense," an article he wrote for the April 1937 issue of the *Schweizer Monatshefte*. He described the country as being caught between a north-south fascist axis and an east-west communist one. He thought that one of the prime symptoms of the degeneration of the Swiss folk was its feminization due to the spread of technology. He recommended a drastic reorganization of the Swiss school system that would emphasize a traditional classical curriculum and a renewed involvement in the Swiss land. Jung's wartime pronouncements about the cultural crisis facing Europe reflect many of Wili's themes and indicate a familiarity with his writings.

The conservative *Monatshefte* published a review of Jung's *Psychology and Alchemy* in its March 1944 issue. The reviewer was a University of Zurich student Arnold Künzli who had been corresponding with Jung for about a year. He wrote that the God-seeker in Jung had combined with the doctor in him to reanimate the Christian belief system and rescue it from moribund dogmatics. This would be achieved through contact with the living religious impulse found in the unconscious of modern individuals; the resulting individuation process had its analog in the complex symbolism developed by the alchemists. He closed with a reference to Jung's recent appointment to the faculty of the University of Basel hoping that Jung would succeed in leading a society of seekers through the crises at the reeling *universitas*.[20] Künzli's previous review of Jung was of "On the Psychology of the Unconscious" and appeared in *Der Zürcher Student*.[21] This publication was one of the many extreme right-wing periodicals published by the National Front with funds that were sent from Berlin.[22] (Jung was to continue an affiliation with this publication after the war with letters of his appearing in it in 1949 and in 1958.) Künzli's closing reference to Jung's appointment to a chair at his alma mater also has a political subtext since it was reported that leftist students protested Jung's appointment, which he was unable to assume in any case due to ill health.[23]

Künzli aptly referred to Jung's "Paracelsan streak" in his 1944 review. Jung's alchemical studies were greatly influenced by the writings of Paracelsus (1493–1541), the great Swiss alchemist/physician. Jung undoubtedly saw his own career foreshadowed in that of Paracelsus whose stormy temperament and unconventional views alienated many of those around him. Some called him a charlatan while to others he was a sage. Jung gave two talks in honor of the four hundredth anniversary of Paracelsus' death; the following quote is from "Paracelsus the Physician," which he read to the annual meeting of the Society of Nature Research, Basel, September 7, 1941.

> Paracelsus was both a conservative and a revolutionary. He was conservative as regards the basic truths of the Church, and of astrology and alchemy, but sceptical and rebellious, both in practice and theory, where academic medicine was concerned. It is largely to this that he owes his celebrity... I feel that I ought to apologize for the heretical thought that if Paracelsus were alive today, he would undoubtedly be the advocate of all those arts which academic medicine prevents us from taking seriously, such as osteopathy, magnetopathy, iridodiagnosis, faith-healing, dietary manias, etc. If we imagine for a moment the emotions of faculty members at a modern university where there were professors of iridodiagnosis, magnetopathy, and Christian Science, we can understand the outraged feelings of the medical faculty at Basel when Paracelsus burned the classic textbooks of medicine, gave his lectures in German, and, scorning the dignified gown of the doctor, paraded the streets in a workman's smock. The glorious Basel career of "the wild ass of Einsiedeln," as he was called, came to a speedy

> end. The impish impedimenta of the Paracelsan spirit were a bit too much for the respectable doctors of his day.[24]

Künzli had personally experienced the darker side of Jung's Paracelsan streak. In his first review he had pointed out that Jung's Romantic vision often came at the expense of scientific empiricism. This triggered a peevish reply by Jung who wrote "Permit me, therefore, to molest you also with my questions."[25] He went on to browbeat the poor student with a list of all the honorary degrees that he had received as a *scientist* (Jung's italics). In a subsequent letter Jung turned his particular scorn on Heidegger.

> Heidegger's *modus philosophandi* is neurotic through and through and is ultimately rooted in his psychic crankiness. His kindred spirits, close or distant, are sitting in lunatic asylums, some as patients and some as psychiatrists on a philosophical rampage. For all its mistakes the nineteenth century deserves better than to have Heidegger counted as its ultimate representative. Moreover this whole intellectual perversion is a German national institution.[26]

Like his role-model Paracelsus, Jung expressed the stormy side of his personality by hurling insults at his intellectual competitors. With the isolation brought on by the war he was looking at events from the long view of archetypal history where things often appeared dark and strange and one needed a thread to find one's way through the labyrinth.

Mercurius Duplex, Double Talk, and Spies

The thread Jung found was a hermetic one. He gave a two-part lecture "The Spirit Mercurius" at the 1942 Eranos Conference whose theme was "The Hermetic Principle in Mythology, Gnosis, and Alchemy." He explored the post-classical evolution of the figure of Hermes through an analysis of a fairy tale and a consideration of Mercurius' various alchemical manifestations. Mercury had many qualities: changing, ambiguous, shifty, and duplicitous, all of which helped make him the god of thieves and cheats. Mercurius is one important manifestation of what Jung called the Trickster archetype. His most extended treatment of it was in the study of a Native American myth cycle that he later collaborated on with Kerenyi and the American anthropologist Paul Radin. The Trickster represents a state of natural undividedness. Although this state contains shadow qualities tricksters are "not really evil, [but do] the most atrocious things from sheer unconsciousness and unrelatedness."[27] Jung's scholarly interests were always deeply connected to his personal psychology so it is appropriate to see how all this was reflected in Jung's life at this time. Jung linked Mercurius to Saturn in the puer-senex dyad with Saturn's lead acting as a counterweight to Mercurius' quicksilver. This dyad was constellated in his exchange with Künzli: the touchiness and sarcasm that Jung expressed in his letters were quintessentially saturnine in nature.

In a postwar interview Jung responded to increasing public charges that he was anti-Semitic. He tried to explain the distinctions between the Aryan and the Jewish psyche that he had made in "The Present State of Psychotherapy." "Since this article was to be printed in Germany (in 1934) I had to write in a somewhat veiled manner, but to anyone in his senses the meaning should be clear."[28] Obviously, the meaning was *not* clear and became the most notorious pronouncement Jung made, the one most quoted by critics ever since as prime facie evidence that Jung was an anti-Semite. Jung tried to put the best possible spin on things by referring to his "veiled" manner of speaking, but the tone of the article is not one of controlled irony but rather of sarcasm and hostility. Jung was indulging here in mercurial double-talk. Two other examples of this from the period will suffice. To a Jewish writer he wrote in 1934 "Your criticism of my lack of knowledge of things Jewish is quite justified. I don't understand Hebrew."[29] Three years later, he wrote to Hauer, "I myself have treated very many Jews and know their psychology in its deepest recesses, so that I can recognize the relation of their psychology to their religion..."[30] Another example occurred in 1936 on the occasion of Jung's participation in the Harvard Tercentenary. To a *New York Times* reporter he said "I am convinced that here is a strong man, a man who is really great."[31] Two weeks later in London he said "Make no mistake about it, he [Roosevelt] is a force—a man of superior and impenetrable mind, but perfectly ruthless, a highly versatile mind that you cannot foresee. He has the most amazing power complex, the Mussolini substance, the stuff of a dictator absolutely."[32] Contradictory statements like these came back to haunt Jung after the war when allegations of anti-Semitic and pro-Nazi leanings brought him unwanted publicity.

Jung linked Mercurius to Wotan but a more plausible case can be made that the Germanic god most closely related to Mercurius was Loki, the trickster god par excellence of Germanic mythology. Loki was constantly causing mischief for the other gods and goddesses; his most reprehensible crime was causing the death of the fair god Balder. He plotted with the forces of darkness and would eventually lead them in their successful assault on Asgard, the home of the gods. Jung made one reference to Loki while amplifying a mandala he had painted that appears as Figure 28 in his article "Concerning Mandala Symbolism." In it an upper figure of an old man in a yoga position [Philemon] is paired with a lower one who is Loki or Hephaestus with red flaming hair. "The old man corresponds to the archetype of meaning, or of the spirit, and the dark chthonic figure is the opposite of the wise Old Man, namely the magical (and sometimes destructive) Luciferian element."[33] Jung's fascination with the Wotanic upsurge in Germany had initially blinded him to the symbolic fact that it was Loki with his duplicity, criminality, and destruction of the social order who better captured the essence of Hitler and the Nazi movement than did Wotan.

Jung would slowly develop this alternative mythological reading of the German phenomenon. In the closing sentence of his Mercurius article he wrote "Lucifer, who could have brought light, becomes the father of lies whose voice in our time, supported by press and radio, revels in orgies of propaganda and leads untold millions to ruin."[34] Besides Lucifer, Jung also associated Hitler with the Antichrist. But in addition to his mythological analysis, Jung made a series of informed psychiatric observations that have often gone unnoticed.[35] In1945 he wrote to an Israeli reporter that Hitler was a hysteric suffering from *pseudologia phantastica*, in other words, that he was a "pathological liar."[36]

Those who claim that Jung only became critical of Hitler after the fall of Germany would find his observations in his prewar English-language interviews enlightening. The Knickerbocker interview gained a lot of attention and led to his being interviewed by Howard L. Philip at the home of E.A. Bennet, apparently at the time of Jung's trip to the United Kingdom for the Oxford Conference. He continued to talk about Hitler as a mystic medicine man type of leader who was guided by his Voice. He was open to the intuitive hunches coming from the unconscious, one that uncannily mirrored that of the German people. This familiar line of analysis was augmented by more specific observations about Hitler's behavior.

> Hitler has never gained a healthy relationship to this female figure, which I call the anima. The result is that he is possessed by it. Instead of being truly creative he is consequently destructive. This is one reason why Hitler is dangerous; he does not possess within himself the seeds of true harmony... He will turn around and say something quite different from what he has said before. He will lose his job when he loses his voice. This might happen, but I do not think it will. Nor do I think he will turn into a normal human being. He will probably die in his job.[37]

This proved to be a fairly accurate assessment of Hitler's psychological development after 1938.

Jung's impression of Hitler during Mussolini's state visit in September 1937 was decidedly more negative than anything he had said previously "[Hitler was] a sort of scaffolding of wood covered with cloth, an automaton with a mask, like a robot... "[38] Another important source of information, especially in light of his comments about Hitler's relationship to the feminine, were the conversations he had with Ernst Hanfstaengl, Hitler's foreign press secretary who had fled Germany earlier that year.[39] Hanfstaengl (1887–1975) grew up in a Munich family deeply involved in the art world. The fact that his mother was American-born led to his attendance at Harvard from which he graduated in 1909. In the early 1920s his family became one of the first respectable ones to open its door to a rabble-rousing demagogue named Adolf Hitler and it was to the Hanfstaengl country house that he fled after the failure of the Beer Hall Putsch. Ernst, affectionately named "Putzi," became an intimate of Hitler and his piano-playing and jokes brought an element of levity to Hitler's inner circle. After 1933 he became increasingly worried about the boorishness and criminality of the top Nazi leadership. Learning that a trip to civil war-divided Spain was to be a pretext for his murder, he fled to Switzerland. He went on to London where he was incarcerated as an enemy alien. Hanfstaengl's intimate knowledge of Hitler and his inner circle was his ticket out of his Canadian internment. He was freed shortly after offering his services as a political/psychological advisor to Franklin D. Roosevelt (Harvard '04). He was put up in a secure house outside of Washington, D.C. where he prepared weekly reports for the president and briefed military and State Department personnel about the Nazi leadership.[40]

Hanfstaengl's insights into Hitler and the Nazis found their way into the wartime report that Walter Langer prepared for "Wild Bill" Donovan of the Office of Strategic Services (the forerunner of the CIA) that was only published in 1972.[41] What is generally not known is that Langer incorporated without acknowledgment another report that had just been prepared by Henry Murray.[42] Besides professional opportunism, another possible reason for Langer's refusal to cite Murray was the latter's affiliation with Jung. That Langer, who had been on the staff of Murray's Harvard Psychological Clinic, never acknowledged this source can probably be attributed to his Freudian bias. This bias was shared by Robert Waite who wrote the afterword to *The Mind of Adolf Hitler* and later made no mention of Jung in his book *The Psychopathic God: Adolf Hitler.*[43]

Jung's psychological insights into Nazi Germany also found their way to the attention of American policymakers through the efforts of Mary Bancroft, one of the most colorful and independent of the Americans who had begun to gather around Jung in the 1930s. A dynamic extravert from a prominent Boston family, she came to Switzerland with her second husband, a businessman. Reading *Modern Man in Search of a Soul* led her to attend Jung's ETH lectures and then his Zarathustra Seminar. Inevitably she began analysis, first with Toni Wolff, and later with Jung himself.

In November 1942 Allen Dulles arrived in Bern to set up the OSS intelligence operation. He crossed the border from France just hours before the Germans occupied the Vichy-controlled part of the country in the wake of the Allied invasion of North Africa. With Switzerland now completely surrounded by Axis armies, Dulles was unable to bring in his trained support staff and so was forced to recruit from among the American expatriate community. Fluent in French and German, Bancroft began analyzing the speeches of the Nazi leaders using a Jungian vocabulary.[44] She and Dulles soon began to have an affair, which did not stop her from becoming a friend of Dulles' wife Clover. The two women shared an enthusiasm for Jungian psychology with the result that one of Dulles daughters eventually became a Jungian analyst. (All this in spite of Dulles' mock protest that "I don't want to go down in history as a footnote to a case of Jung's!"[45])

Dulles soon gave Bancroft her most important assignment, the translation of a manuscript by Hans Bernard Gisevius, Dulles' main contact with the German Resistance. Besides keeping her occupied, it was an opportunity for Dulles to learn more about developments in Germany. Bancroft turned to Jung for advice as to how to proceed. Since Gisevius had read Jung's "Wotan" and felt that it coincided with some of his own formulations about what had been going on in Germany, a meeting of the two was arranged. She later recalled that Jung

> told me that Gisevius and I were going to have an interesting experience working together because we were exactly the same psychological type [extraverted intuitive]. He warned me that if I wanted Gisevius to "spill," I must never ask him for a "fact." If I did, his reaction would be exactly like my own under the circumstances: He would be thrown off-balance and that would be the end of our freewheeling, associative way of communicating, during which I might be able to learn so much.[46]

Later when she asked Jung if Gisevius was close to the edge of being homosexual, Jung exclaimed, "Close to the edge? He *is* the edge!"[47]

When the war started Gisevius (1904–1974) was serving in the *Abwehr* (Military Intelligence Service), which had become the center for anti-Hitler intrigue. Its head Wilhelm Canaris and his former deputy Hans Oster played an important role in organizing the opposition of conservative military men to Hitler. This effort culminated in the unsuccessful bomb plot against Hitler's life on July 20, 1944. Prior to this Canaris assigned Gisevius to Zurich as the German vice-consul as a cover for establishing contacts with Anglo-American intelligence agencies.

What Gisevius wrote in the foreword of his book *To the Bitter End* (1947) expressed in the simplest possible words the philosophy of this entire group of conspirators "I formerly stood on the right... I have not abandoned my conservative views... [and] I hope that new conservative forces will in time arise."[48] This group shared an aristocratic disdain for the Nazis that stemmed from the elitism promoted by the Stefan George Circle as well as from their traditional class sensibilities. This disdain now gave way to dread as they contemplated Germany's impending defeat. They imagined that after Hitler was eliminated they would be able to negotiate a separate armistice with the Anglo-Americans that would lead to their joining Germany in an anticommunist crusade against the Soviet Union (a war aim that, we should remember, Jung did not find unattractive). Living inside the cocoon that was Nazi Germany these men were unable to comprehend the Anglo-American commitment to the policy of "unconditional surrender" that Roosevelt and Churchill had announced as their non-negotiable war aim in consultation with Stalin.

Although officially required to support this policy of total military victory, Dulles operated with a different vision of his mandate. Gisevius wrote "Dulles was the first [Allied] intelligence officer who had the courage to extend his activities to the political aspects of the war. Everyone breathed easier; at last a man had been found with whom it was possible to discuss the contradictory complex of problems emerging from Hitler's war."[49] Gisevius' enthusiasm stemmed from whatever he had heard about Dulles' meetings with Max Egon von Hohenlohe, a German businessman sent by the SS to Switzerland in the winter of 1942. The choice of contacts was an appropriate one since Dulles was well-connected with the highest levels of German banking and business through his work with the New York law firm of Sullivan and Cromwell. He was brought into the firm by his brother John Foster Dulles who had become a partner specializing in the legal intricacies of the reparations payments, an issue that dominated international finance throughout the 1920s. The firm soon represented Germany's leading cartels and smoothed the way for American corporate investment in Germany. Allen brought to the firm his extensive network of government and international contacts gained from his career as a Foreign Service officer that had begun with his attendance at the Paris Peace Conference in 1919.

Dulles became the focus of initiatives by German businessmen to establish contact with Allied governments. Among those businessmen was Georg von Schnitzler who was on the board of directors of IG Farben.[50] IG Farben was deeply involved in the German war economy and

> appears to be the first company to fully integrate concentration camp labor into modern industrial production, and it eventually became known in Germany as a model enterprise for this new technique. Farben executives even provided advice and training on the large-scale use of forced labor for executives from Volkswagen, Messerschmitt, Heinkel, and other major companies.[51]

After the war von Schnitzler was convicted of "plunder and spoilation" and sentenced to five years in prison.[52]

Another high-level connection Jung had to leading German conservative circles was Ferdinand Sauerbruch. Sauerbruch (1875–1951) was Germany's leading surgeon and had settled in Munich in 1918 where he was appalled by the excesses of the Red Republic (he operated on Count Arco-Valley, who was wounded after assassinating Kurt Eisner, the head of the Republic). He became a friend of Franz von Stuck, arguably Germany's leading Symbolist painter (and a personal favorite of Adolf Hitler's). He had been introduced to von Stuck and the Munich art scene by Erna Hanfstaengl, Ernst's sister.[53] In 1928 he moved to Berlin where he became a chief at the Charite Hospital with which he was affiliated for the rest of his career. Like all those deeply involved in the life of the Third Reich, Sauerbruch glossed over his commitment to the Nazi regime. In November 1933 Sauerbruch was one of a select group of German professors including Martin Heidegger, Eugen Fischer, and Emmanuel Hirsch that publicly pledged its allegiance to the Nazi regime; other.[54] In 1937 he shared the first German National Prize that Hitler started as the German equivalent of the Nobel Prize (Germans were forbidden to accept any Nobel prizes after Hitler became furious that the incarcerated German pacifist Carl Ossietzky had been awarded the Nobel Peace Prize in 1935).

Sauerbruch visited with Jung several times during the war. In her journal, Mary Bancroft wrote the following entry for February 9, 1943.

> After our session this afternoon, Dr. Jung asked me if I'd heard that he flew regularly to the Fuehrerhauptquartier to "advise" Hitler...He thinks that the rumor about his flying to see Hitler was started when Dr. Sauerbruch, the famous Berlin surgeon, first began coming here. They had met on several occasions. Sauerbruch was supposedly treating Hitler. "That was enough for my enemies."[55]

It is possible that the two men had met socially when Sauerbruch was practicing in Zurich prior to and during the first years of World War I; they were also both friendly with Friedrich von Müller, Jung's professor in Basel and a colleague of Sauerbruch's in Munich. As a government-approved figure, Sauerbruch would have had little difficulty arranging trips to Switzerland during World War II. The most likely official reason given for his trip would have been his participation in the celebration of the four hundredth anniversary of the death of Paracelsus held at Einsedeln from October 4 to 6, 1941. He had been a member of the executive board of the Paracelsus Society that published the *Acta Paracelsica* (Munich) in five volumes from 1930 to 1932; other members of the society included Bernard Aschner, Alfred Baeumler, Edgar Dacque, Erwin Guido Kolbenheyer, Alfons Paquet, and Karl Wolfskehl. In 1944 a Swiss Paracelsus Society was founded that published the *Nova Acta Paracelsica* (Basel) intermittently until 1948.[56]

Sauerbruch's visit was featured on a front page story run by *The New York Post* on August 15, 1941. The headline reads "Report Hitler in Collapse, Swiss Hear He Has Left Front." The article says that

> Following persistent rumors about the state of Hitler's health was the secret arrival in Switzerland recently of Hitler's personal physician, Dr. Sauerbruch,

> who held conferences with Dr. Tonnis, the brain specialist, and with Prof. Jung, the celebrated psychologist, according to these sources. It was stated that Dr. Sauerbruch disclosed to Prof. Jung the rapid deterioration of Hitler's mental condition.

It obviously became an immediate and well-known secret that the two men were discussing more than Paracelsus. What can be inferred is that the two men were operating within the loosely organized conservative oppositional group that was beginning to take on a more definite outline. They had the professional stature to certify that Hitler's physical and mental conditions had declined to the point that a change in leadership was legally justified. Like most of the group's other initiatives, this one never got past the discussion stage.

As Germany's chief military surgeon, Sauerbruch was entrusted with the life of Claus von Stauffenberg (1907–1944) whose body had been shattered in combat in North Africa in April 1943. After his recovery, he became the dynamic catalyst for action against Hitler. Although not an active participant, Sauerbruch became friendly with the conspirators and allowed them to use his house for some of their secret meetings.[57] After the plot failed Sauerbruch was questioned by Ernst Kaltenbrunnen, the chief of Nazi Security, but was able to talk his way out of arrest and the certain execution that would have followed.

Jung's final personal link to conservative oppositional circles in Germany was Wilhelm Bitter. In 1934 Bitter switched from political science to medicine at the University of Berlin. He trained at the Charite Hospital (where Sauerbruch was on the faculty) and at the Göring Institute where he did a training analysis with the Jungian Kate Bügler. By 1943 Bitter had become involved with the feelers put out by Himmler and the SS to the Anglo-American intelligence services in Switzerland. Through his mentor Max de Crinis, Bitter became acquainted with Walter Schellenberg, Himmler's foreign intelligence chief who discussed with him the Bolshevik threat and the importance of enlisting the Western Allies in a united front against it. He went to Switzerland where he consulted with Jung, who in turn spoke to Carl Burckhardt and Albert Oeri. Upon his return he met with Himmler and came under suspicion for being a defeatist. Bitter obtained a letter from de Crinis authorizing a trip to Switzerland to receive colon treatment and wisely stayed there until 1947. When he returned to Germany he settled in Stuttgart and founded a Jungian-oriented training institute.[58]

Bancroft, Dulles, and Jung shared an interest in the psychological impact of Nazi and Allied propaganda efforts.

> Allen would debrief her on her latest assessments and they would argue the substance: "Hitler's got his facts all wrong," Allen the Jungian man, would huff, provoking Mary, the Jungian woman, to "attempt" to "enlighten" my new boss about the Nazi theory of propaganda, how it had nothing to do with presenting

> the facts accurately but solely with an appeal to the emotions of the German people.[59]

She noted that Dulles was always interested in Jung's opinion of the effectiveness of Allied propaganda. Jung reiterated the futility of negative propaganda: why would a German risk his life to listen to a forbidden broadcast if it was only going to scold him?[60]

The Allied landing at Normandy on June 6, 1944, was the beginning of the end of the Third Reich. Allied troops reached the Swiss border on August 23 and Paris was liberated two days later. German troops were pushed out of eastern France in the autumn and their efforts to regain the offensive in December were thwarted at the Battle of the Bulge. With the capture of the bridge at Remagen on March 7, 1945, the Allies were poised to enter Germany.

Eisenhower had issued a series of proclamations to prepare the population for Germany's impending defeat and occupation. The fact that they impressed Jung prompted Dulles to solicit a letter from him that could be forwarded to Eisenhower. After Bancroft proofread the letter and made some suggestions it was sent with an accompanying letter from Dulles. Jung's letter of February 1 read in part

> These proclamations, couched in simple, human language which anyone can understand, offer the German people something they can cling to and tend to strengthen any belief which may exist in the justice and humanity of the Americans. Thus they appeal to the best in the German people, in their belief in idealism, truth, and decency.[61]

Two days later Dulles responded "I am leaving for Paris on Wednesday to be gone a few days, and I shall take your letter, in which I know General Eisenhower will be much interested. I shall also pass on messages from you to Paul Mellon."[62] Mellon was serving in the army, stationed in London. He was interested in visiting Jung to get advice about German national psychology to help with propaganda efforts. His commanding officer contacted Dulles who gave the "OK" but it all came to nothing when Germany surrendered on May 7 before Mellon could make the trip.[63]

Jung began to have reservations about Hitler's intentions before the war broke out. As the fortunes of war shifted he began to align himself with conservative groups inside and outside Germany that were anticipating the next stage of global power politics. The wartime collaboration between the United States and the USSR was soon to turn to Cold War competition. With his professional credentials and new personal connection to Allen Dulles (who would become the director of the CIA in 1953), Jung became a charter member Cold Warrior while deflecting highly publicized accusations that he was anti-Semitic and a Nazi sympathizer.

Chapter 7

The Cold War Years

As the war in Europe came to an end, the controversy over Jung's conduct and statements during the Nazi era began. By the 1950s the controversy had subsided and Jung entered the last creative phase of his life. Besides the publication of his major study of alchemy *Mysterium Coniunctionis* this period saw the growing institutionalization of his psychology with the opening of Jungian training centers in various European and American cities. He continued his role as a public intellectual by publishing works for the general public, the best-known being *The Undiscovered Self* (1957). In it he presented his critique of the collectivist trend in modern society, especially in its Soviet edition. His strong anticommunist stance provided a hook for his critics who contrasted it to his sympathetic interpretation of Nazism before the war. Jung followers were busy in the first years after the war countering the allegations that he had been anti-Semitic and a Nazi sympathizer. This controversy subsided but still appears in the literature about him.

Peace Arrives

On May 11, 1945, several days after Germany's surrender, *Die Weltwoche* of Zurich published its peace issue that contained an interview with Jung entitled "Will the Souls Find Peace?"[1] In it Jung articulated the observations on the German situation that he would elaborate further in other articles over the next year. His most controversial comment was that, psychologically speaking, all Germans had to admit that they were guilty of the atrocities committed during the war. The psychologist "ought not to make the popular sentimental distinction between Nazis and opponents of the regime."[2] He offered as proof material from two of his current patients, both anti-Nazis, whose dreams displayed "the most pronounced Nazi psychology" with all its violence and savagery.

This position raises several troubling points. First is the fact that it was a violation of the confidentiality essential to the therapeutic relationship. One can only imagine the impact this public revelation had on the course of their analytical work with Jung. Second, it overemphasized the collective aspect of the psyche at the expense of the consciously achieved individual point-of-view that Jung always made his therapeutic goal. Finally, his willingness to dismiss the distinction between a Nazi and an anti-Nazi who had risked his life to oppose the regime as "sentimental" merely demonstrated the real distance Jung had from the moral choices people had to make in Nazi Germany. Jung's blurring of the two also had a self-serving motive. In the wake of Germany's overwhelming defeat, it was a convenient way for him to rationalize his compromising views of the Nazi phenomenon. Much of what Jung published during this period was written with this in mind.

Jung's observation was reminiscent of the one he made after World War I that he had noticed an activation in the unconscious of his German patients that was characterized by primitive aggression. It is distressing that Jung failed to see similar signs when the Nazis came to power. Considering the medical persona that he consistently maintained, one can only wonder what effect a public discussion of this aggressive component by Jung might have had in Germany *before* the war. How differently he would be viewed if his "warning voice" had been raised about this danger rather than the supposed threat posed by Freudian psychoanalysis. Fascinated by the Wotanic upsurge he saw evident in the Nazi phenomenon, Jung had turned a blind eye on his own original insight into what was brewing there. As we have seen he had, in fact, recommended a German attack on the USSR before the war and condoned it a year after it occurred, an act of aggression that brought death to millions.

Toward the end of the interview Jung concluded a discussion of the negative effects of collectivization by turning from the case of Germany to the situation of the Allied victors. "'General suggestibility' plays a tremendous role in America today, and how much the Russians are already fascinated by the devil of power can easily be seen from the latest events, which must dampen our peace jubilations a bit."[3] Something striking about the language in this interview is the degree to which Jung adopted theological terminology to express himself. He made frequent references to "the devil" and "demons." Sometimes he explained them in terms of his model of the psyche while at others he employed them rhetorically. For example, "Now that the angel of history has abandoned the Germans, the demons [the Devil] will seek a new victim." This concern for theological issues had begun during the war and continued in such important postwar works as *Aion* (1951) and *Answer to Job* (1952). In them he presented his interpretation of the Judeo-Christian-alchemical evolution of the God image. His most provocative thesis was that the orthodox tradition had established a one-sided image of God that left no place for his dark side so that over time it became constellated in the figure of Satan. As a corollary of this Jung held that the Augustinian explanation of evil as a *privatio boni* ("absence of good") was inadequate and did not do justice to its psychological reality.

The war prompted Jung to reassess his opinion of the Christian legacy of European civilization and individuals. His assessment was more positive than it had been before the war. Starting in 1918 he had generally seen the activation of the barbarian elements in the personality as a generally healthy development. Now in 1945, he began to publicly consider the consequences of this process. In the *Weltwoche* interview he stated that an admission of guilt was the sine qua non of the moral reeducation of the German people. This could only be achieved through the effort of the individual and not through suggestion.

> The power of the demons is immense, and the most modern media of mass suggestion—press, radio, film, etc.—are at their service. But Christianity, too, was able to hold its own against an overwhelming adversary not by propaganda and mass conversions—that came later and was of little value—but by persuasion from man to man.[4]

Jung's words seem more pious than heartfelt when we remember that in his Wotan essay he made a snide remark about the Confessing Church then undergoing persecution while sympathizing with Hauer's German Faith Movement. This opinion was also consistent with his 1923 letter to Oscar Schmitz that talked about the importance of people returning to their barbarian roots to have a new experience of God (i.e., Wotan).[5]

In 1945 these views were impolitic and Jung began a defensive strategy of revision and self-justification that many critics noticed at the time. Indicative of this shift was Jung's choice of his newest intellectual confidant. Jung depended on an intellectual alter ego to help him formulate his ideas. Freud, then Count Keyserling in 1920s and Hauer in the 1930s each played this role. After the war Hauer was a persona non grata under investigation by French occupation authorities for his Nazi affiliations. An English Dominican priest Father Victor White replaced Hauer as Jung's sounding board for his religious speculations. They quickly grew close and White was invited to speak at the 1947 Eranos Conference. Jung's letters to White are filled with his efforts to explain his psychological forays into theology but their relationship began to cool after White criticized Jung in 1949 for his "quasi-Manichaean dualism," which misunderstood the doctrine of the *privatio boni*.[6]

"After the Catastrophe" and "Answer to Job"

The closest that Jung ever came to a public admission of guilt was in the opening paragraphs of his article "After the Catastrophe," which appeared in the June 1945 issue of the *Neue Schweizer Rundschau*.

> I found myself faced with the task of steering between Scylla and Charybdis, and—as is usual on such a voyage—stopping my ears to one side of my being and lashing the other to the mast. I must confess that no article has ever given me so much trouble, from a moral as well as a human point of view. I had not realized how much I myself was affected... The innermost identity or *participation mystique* with events in Germany has caused me to experience afresh how painfully wide is the scope of the psychological concept of collective guilt.[7]

The first thing to notice is that besides its clumsiness, Jung's classical allusion involved a slip. Jung confused Odysseus' maneuvering past the twin threats of Scylla and Charybdis with the hero's preceding encounter with the sirens. This slip should be taken as a complex indicator of an area of unconscious vulnerability. I have previously interpreted this as being connected to the fact that the sirens were negative anima figures that in turn point to Hitler.[8] In the 1930s Jung had often emphasized the power of the German dictator's Voice, comparing it to the Sybil or Delphic Oracle, and warned that he was dangerous because of his negative anima possession.

We now come to another instance of inaccuracy in the Hull translation. It involves the second sentence of the quote that should more accurately read "I will not conceal it from the reader: never has an article cost me such moral, even human pain." Again, it is more than a matter of quibbling about a translator's stylistic choice of words. Rather, Hull blunts the impact of the deeply personal nature of Jung's revelation by eliminating the references to "pain," "the reader," and to what the writing "cost" him. How Hull came to be chosen by the Bollingen Foundation as the translator for Jung's Collected Works will be discussed later in the chapter. Jung went on to discuss Faust ("so infinitely German") in relation to the German situation. "We never get the impression that he had real insight or suffers genuine remorse. His avowed and unavowed worship of success stands in the way of any moral reflection throughout, obscuring the ethical conflict, so that Faust's moral personality remains misty."[9] One can only wonder the degree to which Faust was serving as a mirror for Jung's own self-examination.

It is striking that Thomas Mann was at the very same time also using the Faust figure to express his understanding of what had happened in Germany; his novel *Doctor Faustus* appeared in 1947. Mann was also thinking about Jung. In a letter to Karl Kerenyi on September 15, 1946, Mann criticized Jung (although Jung's name was deleted in the text by the editor, he was identified in a footnote) by writing of "his odious pro-Nazi pronouncements of 1933." He went to say of Jung that "*not* to recognize immediately such infernal garbage as German National Socialism for what it was, but rather to speak of it at the start in quite different, most distressing terms, was, I think—less excusable, though I find it decidedly tiresome to keep bringing this up against that great scholar."[10]

The greater part of "After the Catastrophe" deals with Jung's elaboration of the psychological consequences of collective guilt on the Germans and the rest of the European community. He reiterated the conservative critique that he had consistently espoused since the 1920s when he declared that "the German catastrophe was only one crisis in the general European sickness."[11] This sickness had its origin in the formation of a large population of uprooted, urban masses in the wake of industrialization.

> What is wrong with our art, that most delicate of all instruments for reflecting the national psyche [i.e., Volksseele]? How are we to explain [deleted: the widespread domination of"] the blatantly pathological element in modern painting? Atonal music? The far-reaching influence of Joyce's fathomless *Ulysses*? Here we have the germ of what was to become a political reality in Germany.[12]

At one point in the article Jung referred to the Reichstag fire in 1933 as a clear signal as to where the incendiary evil dwelt while "we ourselves were securely entrenched in the opposite camp."[13] Even the most sympathetic reviewer of what Jung said in 1933 must acknowledge that this simply wasn't so. Jung, in fact, contradicts himself later in the article when he gave a candid assessment of his opinion at that time when he wrote

> At that time [1933–1934], in Germany as well as in Italy, there were not a few things that appeared plausible and seemed to speak in favour of the regime...after the stagnation and decay of the post-war years, the refreshing wind that blew through the two countries was a tempting sign of hope.[14]

This is a remarkably honest, if not politically correct, thing for a European conservative to say in 1945. He did state for the record that his reservations began in 1937 with his visit to Berlin and his personal observations of the two dictators.

His observations led him to make the psychiatric diagnosis that Hitler was suffering from a form of hysteria called *pseudologia phantastica* in which the person believes their own lies. Jung had made the same diagnosis in discussions with Knickerbocker,but did not then publicize it, suggesting that he did not really have the courage of his convictions. He makes the point in the article that "Hysteria is never cured by hushing up the truth."[15] This is, however, exactly what Jung did before the war. That he needed to feel the weight of public opinion behind him is also evident in the comments he made in the closing paragraphs of 1934 "Rejoinder to Dr. Bally" about "the Jewish problem."

> I have tabled the Jewish question... the first rule of psychotherapy is to talk in the greatest detail about all the things that are the most ticklish and dangerous and most misunderstood. The Jewish problem is a regular complex, a festering wound, and no responsible doctor can bring himself to apply methods of medical hush-hush in this matter.[16]

"After the Catastrophe," "Wotan," and several shorter pieces were published together in the United Kingdom as *Essays on Contemporary Events* (1947), a year after they were first published in Switzerland. In part, this publication functioned as a public relations effort to counter Jung's critics by presenting his side of things. The epilogue most clearly conveys this aim by Jung's selection of numerous quotations from his works that date from 1916 to 1937 demonstrating his interpretation of political movements as mass psychoses. Unfortunately, the quotations he provided did not really address the concerns of those critics. At several points he made remarks that indicate that he did not really grasp what they were saying. "It certainly never occurred to me that a time would come when I should be reproached for having said absolutely nothing about these things before 1945..."[17] He was being reproached, of course, not for his silence but for what many took to be a favorable view of Hitler and National Socialism. If he truly wanted to face his critics squarely he should have quoted his 1934 article "The Present State of Psychotherapy" where he wrote

> Has the formidable phenomenon of Nationalism Socialism, on which the whole world gazes with astonished eyes, taught Freud's imitators [parrots] better? Where was that unparalleled tension and energy while as yet no Nationalism Socialism existed? Deep in the Germanic psyche, in a pit that is anything but a garbage-bin of unrealizable infantile wishes and family resentments. A movement that grips a whole nation must have matured in every individual as well.[18]

Jung's insight into the bipolar nature of the archetypes was open to contradictory interpretation.

> It is impossible to make out at the start whether it will prove to be positive or negative. My medical attitude towards such things counseled me to wait, for it is an attitude that allows no hasty judgments, does not always know from the start what is better, and is willing to give things a "fair trial."[19]

This means that although Jung says that he was certain that with Hitler's coming to power Germany was undergoing a mass psychosis, he chose on medical grounds to "give it a chance." This would be consistent with his observation in his 1932 lecture where he said "There are times in world history—and our own time may be one of them—when good must stand aside, so that anything destined to be better first appears in evil form."[20] His views of the Nazi takeover were influenced by his idealistic, conservative friends who pleaded that abuses were customary in any great revolution. Furthermore, as a Swiss he felt bound to Germany by ties of blood, language, and friendship and wanted to do everything in his power to prevent those cultural bonds from being broken. When Jung observed Germany it was not with the eyes of a doctor, his constant reference to his neutral medical persona notwithstanding, but with the opinions and prejudices common to European conservatives who initially believed that the Nazis would bring about Germany's renewal.

Along with the allegation of pro-Nazism, Jung also had to fend off the charges that he was anti-Semitic. Many critics have faulted Jung for not having written a sustained and personal assessment of the Holocaust. He did make references to Buchenwald and mass exterminations in "After the Catastrophe" but described these as examples of the systemic nature of Nazi brutality rather than as aspects of a calculated policy against Jews. He wrote a passage on "the Jewish problem" that was intended for the epilogue but was never published. He discussed this with Michael Fordham in correspondence in the spring of 1946 in anticipation of the English publication of *Essays on Contemporary Events*. Jung had sent the passage to Gerhard Adler who recommended that it not be included for an English public unaware of the polemics against Jung. Fordham concurred with this opinion and Jung acquiesced in their advice.[21]

What Jung had written, "Remarks of C. G. Jung on His Position on Anti-Semitism," is in the Jung Archive at the ETH in Zurich.[22] It consists of his version of the how the charges of anti-Semitism had plagued his reputation ever since his break with Freud in 1913. He went on to review his reasons for accepting the presidency of the General Medical Society for Psychotherapy in 1933 and the controversy it created. After reading it carefully, one begins to appreciate why Adler and Fordham felt that it was best not to include the piece in the epilogue. It was a personal rehash of things he had said for many years in his own defense. To have published it now would have only hurt his cause by implying that Jung had learned nothing from what had happened (privately he still referred to the topic as "the Jewish problem") and was only interested in justifying his position.

To truly understand *Answer to Job* then, one needs to reconsider it in light of that essay and Jung's tracking of the religious history of Germany. Remember again his 1923 letter to Oscar Schmitz in which he talked about the unique opportunity for people to have a new experience of God; by 1936 he made clear that for Germans this meant the activation of their old storm god Wotan. Blinded by his old prejudices, by Hauer's influence, and by the Klages-inspired scholarship of Martin Ninck he failed for too long to grasp the fundamentally criminal nature of the Nazi regime.

That Jung was dealing with deeply personal issues in *Answer to Job* is evident in something Gerhard Adler wrote to his old friend Erich Neumann, describing the book as "barbaric, infantile, and abysmally unscientific."[23] Many commentators have explained that Jung wrote it as a way to express his lifelong struggle to understand the "dark side" of God and point out that the calamity of World War II lent an urgency to the work. Another observation by Adler connects Jung's motive for writing to his acknowledgment of being affected by the collective guilt of the German people. Adler notes that Jung wrote the book in a state of *Ergriffenheit* ("seizure"), the term Jung had used in his Wotan essay to describe what was happening in Germany. Jung's postwar reflections on the dark side of God should have included an analysis of Wotan but did not. He emphasized his psychiatric reading of events in Germany without mentioning his Wotan hypothesis that viewed developments there from a religious point of view.[24]

The De-Nazification Process

With the war over, the Allied powers established a war crimes tribunal that tried and punished the leading members of the Nazi regime. They also instituted a process of de-Nazification that investigated and meted out punishments to thousands of Germans for their activities during the Nazi era. For intellectuals such as Hauer and Heidegger this meant that they were barred from teaching for a certain period of time. This process of investigation and fall-out impacted other acquaintances of Jung such as Rohan and Schnitzler.

Jung himself had been the subject of reports by the FBI's New York office in September and October 1944. Vague statements had been made alleging that Jung was pro-Nazi and an admirer of Hitler's intuition; there was even a rumor that he was possibly in the United States. An unnamed interviewee (probably the Jungian Eleanor Bertine) rebutted these allegations and that was the end of it. The consensus among Jungians was that this whole mess was due to the animosity of Freudians who never forgave Jung for breaking with Freud in 1913 as well as their hypersensitivity about Jung's acceptance of the presidency of the General Medical Society for Psychotherapy in 1933.[25] In fact, as we shall see, Jung's major postwar critics were not Freudians but leftists who criticized his political stance more than his psychological theory.

Jung's friendship with Allen Dulles now began to pay dividends. The political sympathies of Olga Froebe-Kapteyn, the founder of the Eranos conferences, had come into question so Jung sent her to Allen Dulles who found no merit in the allegations and put the matter to rest.[26] A CIA document dated September 1946 described Eranos as "a forum for anti-Nazi German and Swiss scientists, when Hitler came into power and it became evident that science would be forced to line up with the Nazi ideology" (p. 1).[27] If not written by Dulles it reflected his point of view and gives a slanted version of the conference's original mission and fails to list such participants as Jacob Hauer and Gustav Richard Heyer who were both Nazi Party members. What accounts for this revisionist history? The document continued

> If we want to combat Bolshevism we have to fight it on the ideological level first, The ERANOS contribution is an array of independent thinkers... They had opposed established ideas accepted by the ruling ideology of their epoch which they destroyed. These "destroyers" of authoritative ideas are convincing examples of man's right as individuals with a searching mind, as opposed to an indoctrinated mentality.

It continues, "The ERANOS group is striving to convince them [scientists] that an active participation in political affairs is a supreme obligation of intellectuals in the new social order, for their own security" (p. 3). Clearly, Eranos had to be absolved of any taint of Nazi affiliation in order to draft its participants into the Cold War struggle with the Soviet Union that was just then beginning. This group of proudly "nonpolitical" intellectuals was now portrayed as manning the ramparts of intellectual freedom, a view that suited CIA needs more than it reflected historical facts.

The fall-out from Germany's defeat naturally fell heaviest on Jung's German followers. Barbara Hannah noted that the Psychology Club expelled Heyer and Curtius for each having been "a Nazi during the war."[28] She added that a German colleague told her that Heyer always regretted not having remained loyal to Jung. There are problems with the anecdote. Membership in the Nazi party did not ipso facto mean one was being disloyal to Jung personally or to the integrity of his approach. These men had been Nazis since the 1930s, a fact that had not previously jeopardized their status within the Club. Their expulsion did not stem from any sudden moral epiphany but rather from the need to jettison men whose continued membership would have compromised the Club's reputation. (Remember that in 1944 the Club was more sensitive about Jewish than Nazi membership.)

The evidence shows that Heyer reacted to his repudiation by Jung with anger not regret. In a letter of January 14, 1946, Jung wrote "Heyer had the impertinence to write to me recently that he was only an 'ideologist,' of course no Nazi."[29] Heyer had this to say in a letter he wrote several years later:

> In my personal relationship to Jung I ran quite closely into his "shadow"; for example, in that time he was a passionate supporter of the Nationalist Socialists, when

> it went bad for the regime he cautiously distanced himself and after 45 not only propagated the horrible thesis of collective guilt, but also threw his old German friends and students to the dogs of the denazification powers to eat according to the tried and true motto "catch a thief," an operation which was a full success for him. This and other reasons lead me to speak of a miserable character and, actually, I'm convinced that Jung wouldn't contradict me himself.[30]

After the war Heyer had relocated to Lindau on Lake Constance where he collaborated in a series of seminars with Ernst Speer, a psychiatrist who had been his colleague in the General Medical Society for Psychotherapy.

Jung, similarly, took other German followers to task. In another letter of 1946 he wrote

> Recently a letter burst into my house from Bruno Goetz, the writer [of *Das Reich Ohne Raum*], in which he expressed the wish to visit me immediately. I replied that it was too painful for me to talk to Germans as I had not got over the murder of Europe. Whereupon he drenched me with a flood of literary vituperation. To which I rejoined: Q.E.D. Herr Goetz with his thoughtful answer has once again, but unconsciously, ridden roughshod over the feelings of the non-German in true Teutonic fashion, in order to intoxicate himself with the elation of his noble anger. This is no longer seasonable. The Herrenvolk has become obsolete; the stupendously harmless Herr Goetz still doesn't know that. As a matter of fact he knows nothing at all, and appears mightily justified in his own eyes. I am sorry for these people who have failed to hear the cock crowing for the third time.[31]

Many of Jung's other letters from this time discuss the German situation but in terms very different than those from before the war; the tension is no longer between the civilized and primitive components of the typical German but between "the cultural man and the devil... evil in Germany was rotten. It was a carrion of evil, unimaginably worse than the normal devil."[32]

In spite of the postwar fall-out Jungian psychology did survive in Germany, its center shifting from Berlin to Stuttgart. In 1942 Olga von Koenig-Fachensfeld had become the managing director of the Institute's branch there, joining two other Jungians Jutta von Graevenitz and Wilhelm Laiblin.[33] This laid the groundwork for what came later; Wilhelm Bitter moved there after his stay in Switzerland and founded the Institute for Psychotherapy in 1948. In the same year, he founded the Stuttgart Society for Medicine and Pastoral Care in cooperation with Rudi Daur who had previously been active in the Kongener Kreis. They sponsored a series of annual conferences that included such participants as J. Meinertz, G.R. Heyer, Karlfried Graf von Dürckheim, and Jean Gebser. In 1957, Bitter founded the Stuttgart C.G. Jung Society that was to receive Institute status in 1971. The Hippokrates Verlag survived the war and became an outlet for such Jungians as Gustav Schmaltz.

Jung's Critics in Switzerland and in the United States

Jung published the *Aufsätze zur Zeitgeschichte* as a response to criticism being leveled against him by the leftist press in Switzerland. His 1945 *Weltwoche* interview had prompted *Die Nation* (Bern) and *Vorwärts* (Basel) to write articles contrasting Jung's current opinion of Germany with those he expressed in the 1930s. The *Nation* article referred to Jung's appearance before an enthusiastic audience at the University of Frankfurt in the summer of 1933. In the *Vorwärts*, Erich Kästner lambasted Jung for asserting after the war that there was no distinction between the Nazis and their opponents while going into considerable detail about the differences between German and Jewish psychologies in 1934.

On June 12, 1946, *Die Nation* ran an article "The Political Prognostications of C.G. Jung" by Franz Keller. He stated that Jung's sympathy for authoritarian regimes was well-known in antifascist circles and that he belonged to that class of Swiss who supported the policies of Bundescouncillors Etter and Pilet-Golaz. In this and in other articles he wrote for *Volksrecht* (July 26, 1948, and n.d.), Keller, a Social Democrat, criticized Jung's bourgeois allegiance to an authoritarian democracy supported by these men and first proposed by Plato; leaders were necessary to direct the general population along lines prescribed by the archetypes.

Alex von Muralt, another critic of Jung's, wrote several articles around this time, the lengthiest of which was "C.G. Jung's Position Regarding National Socialism."[34] Less overtly political than Keller, Muralt focused his analysis on "Wotan." Like all of Jung's critics, he contrasted the markedly sympathetic tone of the Wotan article with the critical position he took in "After the Catastrophe." There Jung discussed National Socialism in psychiatric terms rather than in the religious terms that he used in the 1930s. Hitler goes from being a mystic medicine man to a pathological liar. The rhetoric and argument that Jung had employed in "Wotan" indicated, at best, an ambivalent attitude toward National Socialism. Muralt discerned that this was derived from Jung's relativistic *Weltanschauung* and quoted him: "even in the best there is then a seed of evil, and nothing is so bad that some good cannot come from it." Muralt pointed out that Jung had ignored the fact that Wotan was first and foremost a war god who had innumerable devotees in Nazi Germany, the foremost of whom had been Heinrich Himmler.

Muralt also claimed that Jung's 1934 article "The Present State of Psychotherapy" played into Nazi hands by distinguishing between an Aryan and a Jewish psychology. Pointing to the Bally controversy Muralt highlighted two comments by Jung that revealed his ethical shortcomings: "Martyrdom is a singular calling for which one must have a special gift" (CW 10, p. 537); and "To protest is ridiculous—how protest against an avalanche? It is better to look out" (CW 10, p. 538). Muralt suggested that the special gift Jung was lacking was the moral courage to speak out against an inhumane system.

One newspaper *Die Tat* of Zurich defended Jung against the attacks from the Social Democratic press. This paper was affiliated with the Landesring, the political party that sponsored Jung's bid for elected national office back in 1939 and so perhaps the editors felt some loyalty to him. Jung's cause would have been most sympathetically supported by his old colleague Max Rychner who was currently the paper's feuilleton editor.[35] Remember that he was formerly the editor of the *Neue Schweizer Rundschau* and had been responsible for Jung's publishing in that journal in the late 1920s and early 1930s.

The paper also came to his defense in response to a critical article that appeared in New York's German-language newspaper *Aufbau* on December 14, 1945. Its author was W.G. Eliasberg, a founding member of the General Medical Society for Psychotherapy, who had been forced to emigrate by the Nazis. He had nothing but disdain for such former colleagues as Schultz, Schultz-Hencke, Kranefeldt, Häberlin, and Cimbal whom he labeled "Jung-worshippers." He directed his animosity at Jung for distancing himself from his former teacher Freud in 1934. "As it happened—explainable or unexplainable—he reached the climax alienation neatly at the time when it appeared that Nazism had come to stay for the next 1000 or 3000 years." He went on to quote Jung's appraisal of National Socialism in "The Present State of Psychotherapy." He then pointed out that Jung was now in his *Weltwoche* interview disavowing "his most honored unconscious and the ancestral soul" when he compared the German to a drunk waking up from a hangover. Eliasberg ended the article on a sarcastic note by saying "If he should try to come here to his devotees then we will remind him of that and make it clear to him that an archetype as fickle as his can't do any business here. The $200,000 he wormed out of America for his institute is the limit."

The article that triggered criticism of Jung in the United States was a one-page piece "Dr. C.G. Jung and National Socialism" in the September 1945 issue of *The American Journal of Psychiatry*. It contained the quotes from the "Present State of Psychotherapy" and the *Weltwoche* interview that were picked up by Jung's critics. The author S.S. Feldman also gave a short and inaccurate account of Jung's assumption of the presidency of the General Medical Society for Psychotherapy. In particular, he incorrectly stated that Jung collaborated with Matthias Göring on the publication of *Deutsche Seelenheilkunde* in 1934. This and several other inaccuracies (e.g., that Ernst Kretschmer had been forced to resign the presidency because he was Jewish) would be repeated and made the basis of the historical accounts that began to enter the secondary literature.[36]

The public debate about Jung first began in the pages of *The New York Herald Tribune* with letters to the editor published that fall under the alternate headings "pro-Nazi" and "Anti-Nazi." The pro-Nazi letter was written by Albert Parelhoff who would become Jung's most vociferous critic. One anti-Nazi letter was co-signed by Esther Harding and Eleanor Bertine, another was from Carol Baumann who divided her time between New York and Zurich during those years. The debate, framed as a simple dichotomy, was a lopsided one. Jung's defenders did not really meet his critics head-on but rather provided alternative quotes that put Jung in a more flattering light. They also relied on their personal testimonials to Jung's integrity to convince his critics that he was no Nazi. They also pointed out (compliments of Jung but unsubstantiated) that he had been put on a Nazi blacklist and marked for execution if Switzerland were ever invaded.

The Analytical Psychology Club of New York had written to Jung for information to utilize in his defense but in an unpublished letter he responded "It is much better not to mix in with such dirty things. Otherwise you simply pour forth blood into that monster, which, if left alone, would die of its own poison afterwhile [sic]."[37] This in fact would not prove to be the case with the result that the strongest defense of Jung was to come from outside the Jungian camp.

Ernest Harms was a child psychotherapist living in New York where he was editor of the journal *The Nervous Child*. He was Jewish and had been forced to emigrate from Germany in the 1930s where he had been a member of the General Medical Society for Psychotherapy. This first-hand experience gave him a unique perspective on the controversy and informed his article "C.G. Jung—Defender of Freud and the Jews, A Chapter of European Psychiatric History Under the Nazi Yoke" that appeared in the April 1946 issue of *The American Journal of Psychiatry*.

The opening sentence reads "During recent months a wave of misinformation concerning certain periods in the early development of modern analytical psychiatry and psychology has swept through professional periodicals and popular informative literature." What followed was a long, detailed, and generally accurate account of Jung's relationship to Freud and his involvement in the General Medical Society for Psychotherapy. Regarding the history of the Society he made clear a fact routinely ignored by Jung's critics—that he was never a member of the German General Medical Society for Psychotherapy (p. 13). The choice of Jung as president "was motivated by the desire to prevent the whole psychotherapeutic society from falling under the influence of National Socialism. Here again, all subsequent reports regarding attempts to Nazify psychotherapeutic work in Europe are completely false and misleading." Later he continued, "In the true interests of the Jews, it would have been unwise to make a frontal attack on the German psyche, which was seething with hatred. To achieve any result, it was imperative to approach the question rationally and carefully" (p. 16).

Harms' most unique contribution was his contextualization of the remarks Jung had made about Jews and National Socialism in "The State of Psychotherapy Today." He said that he would "quote here in careful translation the pages which have been widely circulated in misleading abbreviations and translations, and from which extracts have been pieced together in a fashion which distorts their meaning" (p. 18). He pointed out that Jung's was a comparative psychology that "does not stop at character and personality differentiation but does go on to typological expressions as they appear in a social, cultural and, finally, anthropological, psychological aspect" (p. 15). He then translated a lengthy passage (now par. 352–356) into English for the first time (and one more faithful to the German than Hull's). "In the accusations made against Jung the following expression in particular has been used as a weapon of attack '...the mighty apparition of national socialism which the whole world watches with astonished eyes...'" (pp. 22–23). Harms went on to clarify Jung's use of "powerful" and "astonished" (which had been mistranslated as "admiring"). Both words were used objectively to describe Nazism as a phenomenon without implying sympathy for it.

Harms went on discuss the 1934 Bad Nauheim conference where Jung gave due credit to Freud in his address "The Theory of the Complexes" for which he was rebuked by the Nazi press. He also explained how Jung was able to successfully implement the changes in rules that allowed for individual (i.e., Jewish) membership in the International Society. He pointed out that no anti-Semite would then have published a book, as Jung did, with a contribution by a Jewish author. Finally, having reminded his readers that many of Jung's most talented and loyal followers were Jewish, he quoted a letter from Jung to one of them. Despite its accuracy Harms' article had little effect on how the controversy developed. It was ignored by both Jung's critics and his New York followers who considered Harms an interloper.[38] In spite of this, he continued to admire Jung, writing an obituary article at the time of his death and leaving his personal papers to the Kristine Mann Library.

In 1946–1947 *The Protestant* ran a three-part story (June–July, August–September, February–March) by Albert Parelhoff whose title "Dr. Carl G. Jung—Nazi Collaborationist" was a blunt rejoinder to Harms. The magazine was not a religious magazine as one might expect from its title, but a far leftist one. A subsequent issue ran a column entitled "Jung's Unclean Hands" that said "It [insanity] may be the atom-bomb madness of Truman's America which is installing the Nazi industrialists and Christian Fascism in Western Germany in preparation for such a war as Hitler could never imagine in his wildest fantasies." Parelhoff himself was remembered by Karl Shapiro as

> a caricature of a Red—slouch hat pulled down over one eye and chewing a cigar, and he kept his hat on. He asked me *sotto voce* if Coleman [Shapiro's chairman

> at Johns Hopkins University] was "all right," which put me on my guard. He showed me papers purporting to prove that Jung was a Nazi . . . [39]

Parellhoff based his attack on the most compromising passages in Jung's works, in particular "The Present State of Psychotherapy," "Wotan," and the Knickerbocker interview. His articles are characterized more by their journalistic bombast than critical analysis ("the so-called *Heilsweg* of Jung had merged with the Heil Hitler! Weg of the Nazis"). He did, however, identify the accommodating tone and moral relativism of Jung's writings. He emphasized how Jung's Wotan thesis justified the Nazi ideology of irrationalism and glorified Hitler's charismatic leadership. His leftist point of view comes across in various passages: "Nazi Germany and Fascist Italy—by official invitation—were well represented at the Harvard Tercentenary. Perhaps the political spirit of the celebration can be better judged when one considers that no scientists from the U.S.S.R. were present."[40] In the final article he wrote "Jung worked in his anti-Russia propaganda but nowhere in the Terry Lectures did he as much as openly name Hitler and the Nazis."[41]

The Parelhoff articles were to be a resource for Robert Hillyer when he initiated the controversy over the awarding of the Bollingen Prize for Poetry to Ezra Pound in 1949. He published articles in the June 11 and 18 issues of *The Saturday Review of Literature* in which he accused the committee appointed by the Library of Congress for being part of a conspiracy whose aim was "the mystical and cultural preparation for a new authoritarianism." Jung's name was dragged in because the award was named after the location of his retreat tower on Lake Zurich and the fact that Jung supposedly shared Pound's fascist sympathies.

Early in the first article "Treason's Strange Fruit," Hillyer recalls

> I had personal contact with Dr. Jung's Nazism. At luncheon during the Harvard Tercentenary of 1936, Dr. Jung, who was seated beside me, deftly introduced the subject of Hitler, developed it with alert warmth, and concluded with the statement that from the high vantage point of Alpine Switzerland Hitler's new order in Germany seemed to offer the one hope of Europe. (P. 19)

This led to a flood of letters pro and con to the *Review* that lasted for months.

The controversy continued with the publication of a pair of articles, one defending and another criticizing Jung, in the July 30 issue under the title "What About Dr. Jung?" Rather than soliciting a contribution from a Jungian loyalist, the editors ran "A Misunderstood Man" by Philip Wylie. He was a syndicated weekly columnist and had used Jung's ideas in his two books *A Generation of Vipers* (1942) and *An Essay on Morals* (1947). They had first met at the Harvard Tercentenary and apparently hit it off since Jung was Wylie's house guest when he gave the Terry Lectures at Yale the following year. Wylie embodied an intellectual type, the iconoclastic social critic, who appealed to Jung and reflected his own self-image (H.G. Wells had been one such in the 1920s and J.B. Priestley would soon be another).

Wylie reminisced:

> In 1936 everyone was talking about Hitler, whose name required no deft introduction then. It is possible that Jung was pulling Hillyer's leg—an act for which he is renowned. But it is more likely that Mr. Hillyer failed to note the exact content and purport of Jung's words. What are the *facts*? In 1936 Jung was (and has been ever since) a vehement antagonist of Russian collectivism. Jung is a German Swiss with a Germanic education. Not unnaturally, he hoped that the German people would find a way to orient themselves against Red aggression and to fend it off. Millions of Americans held the same hope...

He went on to explain that Jung's Wotan hypothesis was intended as an explanation not an endorsement of Nazism. He recalled that Jung had spoken at length about how insane the Nazi leaders were while staying with him in 1937 just a month after his Berlin visit. Wylie was more circumspect regarding Jung's attitudes toward Jews. He began by acknowledging the influence of the animosity Jung harbored toward Freud after their break. He then made a reference to an apocryphal quote of Jung's about the "inferior" nature of the Jewish unconscious and distanced himself from it. This, Wylie argued, was a misinterpretation of the passage in "The Present State of Psychotherapy" where Jung spoke about the higher potential of the Aryan psyche, a view stemming from his energic view of the psychic dynamics and implied no value judgment.

Wylie closed by presenting his credentials to be writing in Jung's defense. "Dr. Jung has written me, after reading my works, that I understand his theories more completely than anyone else writing about them in this country."[42] He continued "[Jung] knows that I have been vehemently opposed to Nazism since I first encountered it in the Twenties—that I have been an articulate foe of Communism since my visit to Russia in the Thirties—and that I am one of the nation's most outspoken foes of anti-Semitism..." To Wylie, Jung's critics had chosen to ignore the lifework of a man dedicated to cultivating individual self-awareness and instead concentrate on a few passages taken out of context to present Jung in the most negative possible light.

The other article was written by Frederic Wertham, a psychiatrist who had first criticized Jung in a 1944 book review in which he called Jung "one of the most important influences on fascist philosophy in Europe."[43] Here he began by saying that Jung "hoisted the swastika banner in a scientific field" by accepting the presidency of the General Medical Society for Psychotherapy.

> The Nazis had a difficult job of finding a psychotherapist or psychoanalyst with a big name. Everyone knew that only Jung would lend himself to such a

> step. For this act was a major political event in the cultural conquest of Central Europe by the Nazis. German psychotherapy had found its Führer...

Jung's example, Wertham argued, influenced many wavering intellectuals into accommodating themselves to the new regime. Wertham emphasized the similarity between Rosenberg's *Myth of the Twentieth Century* and Jung's amalgam of mysticism, occultism, and obscurantism. He went on to say that this double-talk was at the core of Jung's lectures at German universities where it appealed to the irrationalism and nationalism of the student audience. These speeches were of even greater service to the Nazi cause than those of Heidegger. "While millions with the wrong archetype were on their way to death... Nazi writers continued to refer to Jung as 'the great researcher of the soul.'" He concluded with the sarcastic opinion that Pound actually deserved the award since it should have been named the Berchtesgaden (Hitler's mountain retreat) rather than the Bollingen Award.

This line of invective against Jung continued in another article by Wertham, "The Road to Rapallo," published in *The American Journal of Psychotherapy* in October 1949. It was a critical analysis of the Ezra Pound case that questioned the validity of the insanity diagnosis. Inevitably, his passing reference to "Jung the fascist" prompted letters to the editor. Werner Engel, a New York Jungian analyst who was Jewish, wrote in Jung's defense. Responding to Engel, and ignoring Harms, Wertham noted that "not one prominent non-Jungian psychiatrist or psychoanalyst has come out with a clear straight forward defense of their famous living colleague." He went on to refer to Jung's popularity in the re-Nazified circles of Central Europe and concluded with the assertion that Jung could have saved Freud's sisters who had been sent to a death camp. "My article dealt with Ezra Pound, and not with Carl G. Jung. Otherwise, I would have called it not 'The Road to Rapallo,' but 'The Road to Auschwitz.'"

The final installment in this long, messy controversy was the interview of Jung conducted by Carol Baumann and published in the *APC of NYC Bulletin* in December 1949. After four years of negative scrutiny Jung was now ready to reverse his earlier advice that it was "much better not to mix in with such dirty things." To this sympathetic interviewer Jung opened with the statement that all the passages cited by Hillyer had been tampered with out of malice or ignorance. He referred to Harms' article several times, saying that he could add little to what Harms had written. He defended his assumption of the presidency in 1933 as the honest effort of a scientist from a neutral country to keep alive an international organization. To further counter the impression that he was a Nazi collaborator, he cited his successful effort to revise the Society's bylaws to help Jewish colleagues. He also emphasized his friendly relations with such Jewish analysts as Harms and, from his own school, Gerhard Adler and Erich Neumann (whose name replaced Jolande Jacobi in the original transcript).[44]

Jung denied Hillyer's claim that he had expressed admiration for Hitler at the Harvard luncheon, saying that he had always been concerned for the future of Europe. He did reiterate his conservative opinion that in the early years "before the power devil finally took the upper hand" Hitler did bring about many reforms that served the German people constructively. He then referred to his "Wotan" thesis as an apt characterization of what had seized the Germans, stirring up their long buried past. "I wrote this article in 1936 as a warning for those who could understand its implications." He avoided discussing the more troubling statements he had made in it, most notably his admiration for the "enthusiastic" scholarship of Hauer and Ninck and his recommendation that members of the German Christian movement join the German Faith Movement. Many critics took aim at Jung who, along with his defenders, sought to deflect their charges. The controversy slowly died down but would be revived from time to time in the years ahead.

Institutionalization and International Reputation

During this period, Jungian psychology was undergoing a process of institutionalization and increased visibility that was to shape the course of its development down to the present. In 1948 the first training institute for Jungian psychology opened in Zurich with programs in English and German. At the same time, plans were underway for the publication of Jung's Collected Works in a standardized English translation. This was the brainchild of Mary Mellon who had become infatuated with Jung and had gone to Zurich before the war in the company of her husband Paul, heir to the Mellon banking fortune. In 1940 she discussed with Jung her desire to publish both his works and those emanating from the Eranos conferences. It wasn't until after the war that this project was taken up in earnest. Since plans were already underway for the publication of Jung's Collected Works in the United Kingdom by Routledge Kegan Paul the Bollingen Foundation arranged for the American rights.[45]

The editorial board consisted of Herbert Read, Gerhard Adler, Michael Fordham, and William McGuire (executive editor). Over the years most of Jung's major works had appeared in English but the decision was made to retranslate all that he had written. R.F.C. Hull was hired in spite of the fact that he had no previous familiarity with Jung's writings. Jung overcame his initial reservations about Hull and came to appreciate his talents. To be closer to Jung and the Eranos circle Hull eventually moved with his family to Ascona, Switzerland.

The Collected Works, which eventually spanned twenty volumes, were released at a slow but steady rate through the 1950s and into the 1970s. The volume of greatest interest to us was Volume X: *Civilization in Transition*, which was released in 1964 and included most of the articles discussed in this book. That the translations of "The Role of the Unconscious," "The State of Psychotherapy Today," and "Wotan" were a special concern is evident from the editorial correspondence. In April 1963 McGuire and Hull exchanged letters discussing which of several different words would be the best translation of "Neger."[46] As my various retranslations in this book have made obvious, Hull did not adhere to Barrett's request that his translations be completely faithful to the German original. He was responsible for the deletion or mistranslation of dozens of words; in almost every case the original English translations are more accurate if stylistically less polished than Hull's.

The intellectual climate of the postwar period was congenial to Jungian thought.

> The response to economic and imperial decline was in Britain of the forties a literary ambience of despairing resignation, suspicion of and incapacity to sustain an advanced technological society, and an intense but short-lived Christian revival. The leading British writers of the time—T.S. Eliot in poetry and drama, F.R. Leavis in literary criticism and cultural commentary, J.B. Priestley in fiction, Arnold Toynbee in metahistorical speculation—shared this temperament.[47]

Priestley was introduced to Jung by Gerhard Adler in 1946 and gave a BBC radio broadcast on Jung's psychology on June 18 that proved very popular. Jung was impressed by Priestley's summary of his thought and agreed to give a talk on BBC himself. He delivered "The Fight with the Shadow" on November 3 and it became the Introduction to the English edition of *Essays on Contemporary Events.*

These broadcasts along with the founding of the *Journal of Analytical Psychology* increased the visibility of Jungian psychology in the United Kingdom, a situation that did not go unnoticed. While visiting New York, the British literary figure Cyril Connolly was interviewed by *The New Yorker*. "Mr. C. informed us that Jungians are getting dangerous in England, creeping in from all sides."[48] To help counter this perceived threat, he solicited a critique of Jung's theories from the British psychoanalyst Edward Glover for his new magazine *Horizon*. It appeared in numbers 105, 107, and 111 and was later published as *Freud or Jung?* Glover criticized Jung's model of group psychology for giving primary place to a leader who followed his "inner Voice." Glover then quoted things Jung said about Hitler that made it seem that he endorsed the *Führer prinzip*. "These indications of Jung's political orientation and sagacity are embedded in a mass of generalizations from which the contrary impression might appear that his concern had always been with the daemonic (reactionary) aspects of any group expression of [the] Collective Unconscious..."[49]

The reference to Arnold Toynbee reminds us of the interest he took in the Jung's psychology. He had devoted a lifetime of effort to his multivolume *A Study of History* and its 1946 abridgement was a best-seller, helping to make him one of the leading public intellectuals of the time, especially in the United States. He had found Jung ideas helpful in understanding the role of the great world religions in sustaining their respective civilizations. In his 1948 *Civilisation on Trial*, Toynbee acknowledged that Jung's theory of the collective unconscious provided him with the clue to understanding the grand patterns found in world history.[50] His support for Jungian psychology took a more concrete form that year when he became a patron of the newly opened Jung Institute in Zurich. His connection continued with an article "The value of C.G. Jung's work for historians" in *The Journal of Analytical Psychology* (1956: I, 2).

Besides their mutual appreciation for spiritual values, Jung and Toynbee shared a suspicion of mass democracy, preferring the oligarchic rule of an educated elite. Remember that Jung had said that "A decent oligarchy—call it an aristocracy if you like—is the most ideal form of government." Toynbee wrote "An oligarchy with a sense of enlightened self interest is probably the best form of government attainable..."[51] Their ideas resonated with American opinion-makers, especially those of the Eastern Establishment who were looking for orientation in those anxious first years of the nuclear age. Henry Luce, the publisher of *Time*, *Life*, and *Fortune*, had met Toynbee in 1942 and was impressed with his command of international affairs. He featured Toynbee on the March 14, 1947, cover of *Time* and recommended his historical vision as the best framework from which to understand the United States' new position of leadership in the world. In spite of differences of opinion over the special role of the United States, Luce continued to promote Toynbee's work, which insured big book sales and lucrative speaking engagements at numerous American colleges.[52]

Jung was featured in *Time* several times: on July 7, 1952, in its "Personality" column and then in the February 14, 1955, issue. The magazine marked his pending eightieth birthday with an cover story entitled "Exploring the Soul, A Challenge To Freud" and suggested that while Freud was the Columbus of the unconscious, Jung may well be its Magellan. After covering his theory of archetypes and approach to dream symbols, the article summarized his life and professional career.

> One of the most controversial issues about Jung—outside psychiatry—concerns Nazi Germany. Some of his writings about race have been abused by others for racist propaganda. Chiefly because he held the editorship of a German psychoanalytic journal during the Nazi regime (his co-editor at one time was a relative of Hermann Göring), Jung has sometimes been accused of Nazi sympathies. Jung's position: as a foreigner of renown, he merely took the job to insure what he could of German psychiatry.

In April, *Expose*, a New York scandal sheet edited by Lyle Stuart and Paul Krassner, ran a front page story "Time Magazine Honors Nazi Psychiatrist" (issue 40) with a follow-up article in its next issue. The article was signed "Caduceus" but a careful reading makes it clear that it was written by Parelhoff since its content and style are identical to what he had written in *The Protestant.*

The most lasting negative impact on Jung's reputation came as the result of the publication of the second volume of Jones' biography of Freud. *Years of Maturity (1901–1919)* appeared in 1955 and gave canonical status to many unflattering stories about Jung that had circulated for years. The most damaging was a personal anecdote from Jones himself about the 1913 Psychoanalytic Congress in Munich where the final rupture between Freud and Jung occurred. Jung had been reelected president but twenty-two attendees registered their disapproval by abstaining. "He came up to me afterwards, observing that I was one of the dissidents, and with a sour look said: 'I though you were a Christian' (i.e. non-Jew). It sounded an irrelevant remark, but presumably it had some meaning."[53] This along with Freud's reference in "*On the History of the Psycho-Analytic Movement*" to Jung's having put aside "certain racial prejudices" in order to collaborate with him seemed to establish the fact that Jung's anti-Semitism had a long prehistory. While Jones chose to interpret "Christian" as "non-Jew," another reading is not only possible but, given the facts, certain. In *Freud and His Followers* (1976) Paul Roazen writes

> In his autobiography, uncompleted at his death, Jones gave a different and more extended version. "As he [Jung] said good-bye he sneeringly remarked to me: 'I thought you had ethical principles' (an expression he was fond of); my friends interpreted the word 'ethical' here as meaning 'Christian' and therefore as anti-Semitic." Whether it was Jones or his "friends" on Freud's side who made this interpretation, he reported it in his biography of Freud as Jung's literal comment, which by his own later account it obviously was not.[54]

Jung never actually used the word Christian so what seems clear is that his remark was meant to convey his feeling that Jones had acted uncharitably toward him.

The Cold Warrior

In a 1927 article for the *Europäische Revue* Jung had written

> What *does* move more clearly into the foreground is Europe's position midway between the Asiatic East and the Anglo-Saxon—or shall we say American—West.

> Europe now stands between two colossi, both uncouth in their form but implacably opposed to one another in their nature. They are profoundly different not only racially but in their ideals.[55]

A footnote added to the 1959 edition noted that "the East" was now subsumed under the "Russian Empire," which was essentially Asiatic in character in spite of the fact that it reached as far as central Germany. The footnote shows how Jung adjusted his views in light of the Cold War that had developed between the United States and the USSR; it is almost a direct quote of something Wilhelm Bitter had written in his book *Die Krankheit Europas* (*The Sickness of Europe*) about "Asia's outpost against Europe [being] the Russian Empire."[56]

The spirit of wartime cooperation between the two superpowers had quickly evaporated as the Soviet Union began to assert its ideological and territorial designs on Eastern and Southern Europe. In 1947 President Truman's pledge of American support to Greece and Turkey inaugurated a strategy aimed at containing communism around the globe. In addition to military aid the United States began the Marshall Plan that spent billions of dollars on the economic revitalization of war-torn Europe.

That the United States assumed a leadership position at odds with its prewar policy of isolationism was due in no small part to the efforts of the Council on Foreign Relations, an influential group of businessmen, diplomats, and opinion-makers that included Allen Dulles and his brother John Foster Dulles. It promoted its views in the pages of its journal *Foreign Affairs* where George Kennan's article on containment appeared in July 1947. Although the military aspect of containment dominated the public imagination, Dulles saw Soviet ideological aggression against the Free World as the greater threat. "The Russians, by means short of war, will exert themselves to destroy the capitalist system in Europe, and hence to win this particular contest now going on."[57]

It is now well established that U.S. authorities hastily terminated the de-Nazification process in order to enlist German scientists and security personnel into the Western defense system.[58] This policy extended to the ideological struggle as well and had a particular appeal to European conservatives who wanted to distance themselves from their Nazi affiliations and capitalize on their long-standing anticommunist credentials by becoming charter member Cold Warriors; Carl Jung was one of them. The CIA report was part of this strategy as were several articles that Jung contributed to the cultural politics of the Cold War.

In September 1956 he wrote a letter to Melvin J. Lasky the American editor of *Der Monat* (Berlin) that was published the following month as "Wotan and the Pied Piper: Observations of a Depth Psychologist."[59] He was adding his comments to a discussion of this medieval German legend that had extended over the previous issues. He explained the psychological dynamism of the tale using his Wotan thesis and related it to the St. Vitus Dance and beserker phenomena. An editorial note states that *Der Monat* was "an international journal for political and intellectual life, a forum for the open debate of differing voices from Europe, America, and all parts of the world." Besides Jung the October issue had among its other contributors Alberto Moravia, Alistair Cooke, and Walter Laqueur. Lasky is identified in the biographical footnote in the Collected Letters as the editor of the journal from 1948–58 and after 1958 as the co-editor of *Encounter* (London). It has now been established that while he served in these editorial capacities he was on the payroll of the CIA. A cultural affairs operative Michael Josselson activated the Congress for Cultural Freedom, an organization which provided CIA funding to high-quality magazines like *Encounter* in Britain and *Der Monat* in West Germany.[60] On October 19, 1960, Jung wrote a long letter to Lasky expressing his reactions to two articles about Yoga and Zen written by Arthur Koestler for *Encounter*.[61]

Jung's personal connection to Dulles also extended into the 1950s. At this time affairs of state got mixed up with affairs of the heart. Dulles got involved with Claire Booth Luce, wife of Time, Inc. head Henry Luce, who was serving as American ambassador to Italy. Meanwhile, Luce himself had a relationship with Dulles' former mistress Mary Bancroft.[62] Like Henry Murray, the whole bunch may have found some guidance in the example of Carl Jung who was certainly an elder statesman in negotiating the difficult landscape of divided affections. This personal network helps explain the selection of Jung for the cover of *Time* in February 1955. In a 1958 German newspaper interview, the reporter described sitting in Jung's study surrounded by his artwork and book collection. Among the titles he singled out for comment were "several bound volumes of the American political journal *Foreign Affairs*."[63] These volumes were probably gifts from Dulles since it is unlikely that Jung would have gone to the trouble to have them bound.

In December 1956 the United.States. Information Agency broadcast a contribution by Jung to its Voice of America symposium "The Frontiers of Knowledge and Humanity's Hopes for the Future." Jung began his talk with a disclaimer: "I prefer to refrain from incompetent attempts at prophecy, and to present my opinion as the mere desideratum of a psychiatrist living in the second half of the twentieth century."[64] Jung went on to give a clear outline of his theory of schizophrenia based on his fifty years of clinical experience.

In October 1956 two of the Cold War's most serious crises occurred. In Hungary, a popular uprising led to the temporary expulsion of Russian troops from the country. When aid expected from the West did not materialize, Russian forces invaded and crushed the uprising. At the same time British and French forces launched an attack on Egypt in retaliation for Nasser's nationalization of the Suez Canal. Condemned by the United States and the United Nations and threatened with Soviet intervention, the forces were withdrawn.

The fiasco underscored the fact that the two countries were now second-rate world powers. Jung weighed in with his opinions on the crises by contributing to two symposia. "The bloody suppression of the Hungarian people by the Russian army is a vile and abominable crime, to be condemned forthwith." In the second, more reflective piece, Jung focused on the way the crisis was processed in the West. At first Westerners were indignant, but indignation gave way to a moral complacency that ignored the voice of conscience. This voice reminded "the West of those wicked deeds of Machiavellianism, shortsightedness, and stupidity without which the events in Hungary would not have been possible. The focus of the deadly disease lies in Europe."[65] Regarding the Suez Crisis Jung wrote "The Egyptian dictator has by unlawful measures provoked Great Britain and France to a war-like act. This is to be deplored as a relapse into obsolete and barbarous methods of politics." Jung is here taking a tack that was lacking in his prewar recommendation of a German invasion of the USSR. He seems to have found a moral compass that was missing in his previous endorsement of *Realpolitik*.

Jung's main criticism of communism was that it promoted the interests of the collective at the expense of the individual. For him, collectivization was a problem not only behind the Iron Curtain but was an ominous trend in Western society as well. During the 1950s there was widespread concern about the increasing influence of social conformity. Jung had always held "the masses" in low esteem but felt that more than ever people were turning to the state to satisfy their basic needs and give meaning to their lives. He was critical of their efforts to improve their situation through the creation of the modern welfare state. In a tirade in a 1948 letter to Henry Murray Jung referred to the American leader of the United Mine Workers as "ape man Lewis."[66]

Jung's sensibilities were essentially those of a conservative humanist rather than of a liberal humanitarian. His interest was in exploring the psychological basis of human behavior and culture rather than promoting social programs for ameliorating human suffering. This preference can be seen in his sarcastic comment about the missionary work of Albert Schweitzer "who is urgently needed in Europe but prefers to be a touching saviour of savages and to hang his theology on the wall. We have a justification for missionizing only when we have straightened ourselves out here, otherwise we are merely spreading our own disease."[67] In an unpublished memorandum sent to UNESCO Jung described his therapeutic method as being successful only with individuals

> with a certain degree of intelligence and sound sense of morality. A marked lack of education, a low degree of intelligence and a moral defect are prohibitive. As 50 percent of the population are below normal in one or other of these respects, the method could not have any effect on them even under ideal circumstances.[68]

The major text for understanding Jung's social thinking during the Cold War period is *The Undiscovered Self.* Peter Homans correctly analyzes it in terms of Jung's application of his psychological theory to the then-popular theory of mass society as a way to explain the predicament of modernity.[69] Urbanization and the decline of traditional religion created a new mass man in both a sociological and psychological sense. The triumph of a materialistic-rational way of viewing the world deprived people of a connection of a transcendental sense of meaning. Uprooted from their traditional rural way of life, they congregated in large urban centers where they fell prey to political demagogues and the trivial allures of the new consumer society. The answer to the neurosis of modern life and its attendant political and social ills was not a return to traditional religion but an encounter with the psychic depths that have always been the true source of all genuine religious experience.

Peter Homans' broad contextual approach comes up short when addressing the specifics of Jung's familiarity with the theory of mass society. He mentions the influence of Le Bon and Nietzsche on Jung but states that "there is no corresponding debt to the mass-society theorists, who had begun to write in the 1920's and whose work became more and more well known after the Second World War."[70] Jung was, in fact, familiar with two of the three theorists that Homans mentions, namely Max Scheler (1874–1928) and Ortega y Gasset (1883–1955). Jung and Scheler both gave lectures at the 1927 conference of the School of Wisdom and he owned two of Scheler's last works.[71] Although he only mentioned Scheler several times in passing in his own work, it seems obvious that Jung would have heard about Scheler's thesis of mass society from the man himself as well as from mutual intellectual contacts. Jung did not own any of Ortega y Gasset's books including his most famous *The Revolt of the Masses* (1930), but it is certain that the men were familiar with each other's works. They both appeared in a series that also included Scheler and was published by the Neue Schweizer Rundschau Verlag under the editorship of Max Rychner in 1929. Furthermore, they both appeared in the pages of Prince Karl Anton Rohan's *Europäische Revue* and lectured to the Kulturbund. Finally, Ortega y Gasset published a number of Jung's articles in his journal *Revista de Occidente* (Madrid) most of which had first appeared in *Europäische Revue.*[72]

To truly understand *The Undiscovered Self* it is necessary to consider it in its Cold War context. This requires a familiarity with how it came to be written and an appreciation of the Cold War rhetoric that permeates it. Entitled *Gegenwart und Zukunft (Present and Future),* it was published by the *Schweizer Monatshefte* as a supplement to its March 1957 issue. It will be remembered that this was the journal founded by Hans Oehler that originally promoted an anti-Semitic, xenophobic agenda.[73] It survived Oehler's departure and the war, continuing to reflect a deeply conservative perspective that naturally included a bedrock anticommunism.[74]

After an opening paragraph that mentions the Iron Curtain and the hydrogen bomb, Jung writes "Everywhere in the West there are subversive minorities who, sheltered by our humanitarianism and our sense of justice, hold the incendiary torches ready."[75] The only thing capable of stopping the spread of their ideas was intelligence of a portion of the population that he estimated to be about 40 percent of the electorate. He went on to talk about the "army of fanatical missionaries [of communism who] . . . can count on a fifth column who are guaranteed asylum under the laws and constitutions of the Western States."[76] Although he recognized that the United States was the political backbone of Western Europe, he had his doubts about its ability to maintain its position "since her educational system is the most influenced by the scientific *Weltanschauung* with its statistical truths, and her mixed population finds it difficult to strike roots in a soil that is practically without history."[77]

What comes across throughout is the cultural pessimism of a European conservative who had, of course, nothing good to say about Marxism but had little better to say about the liberal alternative, which he considered to be indistinguishable from the Marxist ideal. Suspicious of technical education and utilitarian views of society, he favored a historically based humanistic education that recalled his own upbringing in the Basel of Jacob Burckhardt. He felt that this existed to a greater extent in Europe but that it was threatened by nationalism and skepticism that "both lack the very thing that expresses and grips the whole man, namely, an idea that puts the individual human being in the centre as a measure of all things." The English-language edition published by *The Atlantic Monthly Press* was the result of a conversation between Jung and Dr. Carleton Smith, director of the National Arts Foundation. How they came to meet is as yet not established but one clue is the fact that Smith was a member of the Council on Foreign Affairs so it is probable that the met through the auspices of Allen Dulles and/or Henry Luce.

The book was to become one of the most popular introductions to Jung's thought and was reviewed by Joost Meerloo in the *New York Times* (April 20, 1958) who found it "a passionate plea for individual integrity and for freedom against intrusion." He then managed a back-handed compliment when he wrote that Jung "has traveled a long way from that time in his life when he was infected by the collective mysticism of the Nazi ideology when he postulated a creative Aryan collective unconscious opposed to a destructive Semitic unconscious." On April 26, Robert Graves reviewed it for *The New Statesman* (London) and lambasted it for its banality, illogic, and factual errors. "What I find most unpalatable in this book is a political expediency that condemns Stalin as a monster, for having let three million Russian peasants starve to death, yet makes no direct mention of Hitler's deliberate massacre of over three million Jews." This prompted a letter from Jung's old bête noire Albert

Parelhoff to which Gerhard Adler responded. In turn, Adler's letter elicited one from Hans Keller who concluded,

> I am sure that Mr. Adler will think that I have misquoted Jung. What is more, I am sure he is right. I have found it quite impossible to quote Jung without misquoting him: he always implies the opposite of what he says. "Honesty of attitude?" The question no longer arises. Jung is not dishonest. It is simply that at a certain tragic stage in his brilliant career, the unconscious, which loves contradictions, went to his head and stayed there. (May 24)

The Final Years

After Emma Jung died in 1955, Ruth Bailey an Englishwoman whom Jung had met on his 1925 trip to East Africa moved into his home to provide companionship and care. Although his health was increasingly frail, Jung's mental acuity was evident till the end. An intellectual maverick, his penchant for topics controversial to mainstream scientists was undiminished. He had followed the postwar UFO phenomenon with interest and had collected an extensive file of materials on the subject. He garnered international headlines with his views and published the results of his study *Flying Saucers: A Modern Myth Seen in the Sky* (1958). His main thesis was that UFOs were a living myth that reflected collective anxieties about the hydrogen bomb and overpopulation. Their disc-like appearance indicated a projection of the prime symbol of the self (conscious/unconscious wholeness) onto the skies. He tracked this process through dreams, modern art, and science fiction novels.

Jung's interest led to a spirited discussion with Charles Lindbergh who visited him in the summer of 1959 in the company of his wife Anne and the publisher Kurt Wolff. Lindbergh remembered that he felt "elements of mysticism and greatness about him—even though they may have been mixed, at times, with charlatanism."[78] Jung started to talk about UFOs and seemed to believe all the reports, relying on Donald Keyhoe's book *The Flying Saucer Conspiracy*. When Lindbergh countered that he had discussed the Air Force investigations of sightings with its Chief General Spaatz Jung was not impressed and ended the conversation by saying "There are a great many things going on around this earth that you and General Spaatz don't know about."[79]

The Lindberghs were with Wolff because he had secured the rights to Jung's autobiography for Pantheon Press, which he had founded. The research of Sonu Shamdasani has established that *Memories, Dreams, Reflections* (1989) was not so much an autobiography as a compilation by Aniela Jaffe, Jung's secretary.[80] Along with overseeing this project and advising on the publication of his Collected Works, Jung gave a series of film interviews that brought him to the attention of a new generation of viewers. The interview with John Freeman of the BBC led to the publication of *Man and His Symbols* (1964), the most accessible introduction to Jung's ideas available.

One of the people to whom Jung made himself available in his last years was Miguel Serrano, a Chilean diplomat who had taken an interest in his theory of archetypes. In his *C.G. Jung and Herman Hesse: A Record of Two Friendships* (1966), Serrano reminisced about his encounters with these two old masters. What will come as a surprise to the many readers touched by Jung's ruminations about ultimate things is the fact that Serrano was a committed fascist who views Adolf Hitler as an avatar and used Jung's writings to buttress his argument.[81] Although there is nothing in the book to indicate that Jung knew of Serrano's political sympathies the case does remind us one last time about how a symbolic point of view can be appropriated to justify an esoteric-reactionary ideology.

In the final years of his life Jung developed one of his last friendships with the ETH professor of economics Eugen Böhler (1893–1977) who had become interested in applying a psychological approach to understanding economic behavior. Böhler read *Gegenwart und Zukunft* in manuscript and made suggestions for which Jung thanked him.[82] Their relationship in fact went back years before. Böhler was one of the two faculty members to nominate Jung for appointment to the ETH in 1934.[83] Like Jung he was a contributor to the *Neue Schweizer Rundschau* in the 1930s and so would have been familiar with what Jung was publishing in that journal. In 1940 Böhler coauthored a booklet with Eugen Bircher about the economic and political crisis facing Switzerland; their rhetoric is strikingly similar to what Jung was saying at the same time. It is clear that what drew the two men was their shared conservative economic and political philosophy.

Jung died on June 6, 1961, and Böhler spoke at his funeral. Lengthy obituaries appearing in newspapers around the world focused on his many contributions to psychology, especially his effort to help modern man find his soul (a particularly appealing theme to those concerned with the threat of "godless" communism). Many of them did, however, take note of his involvement in the General Medical Society for Psychotherapy and the controversial remarks he had made about Jews and National Socialism in 1933 and 1934. Implicit in these accounts was an awareness that Jung's life and thought expressed both avant-garde and conservative impulses in often contradictory ways. His Basel education provided the "Archimedean point" from which he viewed the world. His psychology mixed Jacob Burckhardt's conservative hesitation about Progress with Nietzsche's call for self-transformation. Ultimately, Jung sought to balance their disparate views in his theory of individuation.

Conclusion

While tracking down what Jung wrote about politics and race I also researched his professional activities and publishing history, which I learned were either missing from the literature or relegated to footnotes. I wanted to accurately map the German branch of Jung's intellectual and social network and decided to begin with the 1920s since that period was largely terra incognita; I began with his affiliation with Count Keyserling and this led me to Oscar Schmitz and Count Rohan. I continued to connect the dots up until the end of his career and only then circled back to his Basel upbringing and relationship to Freud.

It was only in this later stage that I began to fully appreciate the dynamic tension between Jung's avant-garde and conservative sides. His embrace of a Nietzschean credo is now more evident than ever with the publication of *The Red Book*, which simply does not validate Noll's lurid portrait of the man. Jung was a cosmopolitan intellectual with conservative views on politics and society that became more pronounced as he got older. This is not to say that he ever lost his maverick streak, one that began with a dissertation on spiritualism and ended with a book about UFOs.

Jung shared the stereotypical views of Jews, modernity, and the proletariat popular with the members of his post-Freudian network where his psychoanalytic polemics found a sympathetic audience. In the early stages of my work people would ask, "Well, was he or wasn't he?" At first I would respond as succinctly as possible but finally found myself replying, "Tell me your definition of an anti-Semite and I will tell you the degree to which Jung matches it." Things got interesting when I would point out that some of Jung's most loyal defenders were a group of Jewish followers so a case could be made for his having Zionist sympathies. This apparent contradiction stems from his use of the Romantic concept of the *Volksseele* that postulates a unique psychological orientation for each ethnic group.

Jung can be counted among those intellectuals that Isaiah Berlin identified as belonging to the counter-Enlightenment who upheld the aristocratic principle against the leveling tendencies of modern society. His intention was not to repudiate the Enlightenment so much as to modify its legacy. Burckhardt, Nietzsche, and von Hartmann provided the immediate foundation for Jung's position, which he supplemented with ideas from the pre-Socratic philosopher Heraclitus and the medieval mystic Meister Eckhart. After his discovery of Taoism Jung came to champion an Eastern approach to Enlightenment, one that sought wisdom beyond the scope of intellect. This shift away from Judeo-Christian orthodoxies entailed a departure from its rich ethical tradition as well. His move "beyond good and evil" into the realm of moral relativism is most evident in Jung's pivotal statement that "There are times in the world's history—and our time may be one of them—when good must stand aside, so that anything destined to be better first appears in evil form" (CW 17, p. 185).

Although unnamed, Jung's comment was referring to Nazism, the major political myth of his time. "It is German history that is being lived today…This is real history, this is what really happens to man and has always happened…An incomprehensible fate has seized them, and you cannot say it is right, or it is wrong" (CW 18, p. 164). Much like Gottfried Benn, Jung initially adopted a medical persona that failed to address the criminal reality of the Nazi regime. This was accompanied by a conservative Swiss respect for German *Realpolitik*. Many themes enunciated by German nationalists during his lifetime resonated in Jung; since he failed to heed his own warning voice about the German potential for aggression we find him condoning the German invasion of the Soviet Union a year after it had occurred (a violation, we might remember, of the Hippocratic injunction to "do no harm").

During and after the war he emphasized his Swiss identity while covering his German tracks. He severed contact with his German followers and associates such as Hauer, Rohan, and von Schnitzler who underwent de-Nazification. Jung religious interest turned from Wotan to Job with his writings of this period showing a renewed interest in the Judeo-Christian tradition. Shaken by accusations about his conduct and views before the war he grappled with issues of guilt and atonement. His Gnostic side-stepping around the issue of evil was criticized by Martin Buber and led to his falling out with Fr. Victor White who briefly served as Jung's sounding board.

Concerns about Jung's postwar reputation factored into the decision to publish Jung's Collected Works in a standard English translation by R.F.C. Hull. Hull took many liberties with the text in order to minimize any taint of Nazi sympathies on Jung's part. The writings most subject to sanitation are found in volume ten, one of the last of the series to appear. Biographies have ranged from the reverential to the garish and it is only now that a more accurate portrait of the man is possible. In addition, the

Jungian preference for myth over history has meant that many accounts of his life owe more to storytelling than to scholarship. One example of this genre presents Jung as a sage who during the Nazi era "strayed off the Taoist path" and "fell into darkness." Such an idealization avoids any real engagement with the more problematic, cranky side of Jung's thinking. It seems that sometimes Jung needs to be rescued as much from his admirers as from his detractors.

Notes

1 Basel Upbringing

1. *Memories, Dreams, Reflections* (New York: Vintage Books, 1989) [hereafter *MDR*], p. 3.
2. Ibid., p. 81. He first began to realize this fantasy by building a model of the fortress at Hüningen designed by Vauban. This later influenced the last mandala he did before putting the Red Book aside. See *The Red Book*, ed. Sonu Shamdasani (W.W. Norton: New York, 2009), p. 163.
3. *Dream Analysis Seminar* (Princeton: Princeton University Press, 1984), p. 331. See also *C.G. Jung Speaking* (Princeton: Princeton University Press, 1977) [hereafter *CGJS*], p. 217.
4. *The Collected Works* (Princeton: Princeton University Press) [hereafter, CW], p. 487.
5. Susan Hirsch, "Hodler as Genevois, Hodler as Swiss," in *Hodler, Ferdinand Hodler, Views and Visions* (Zurich: Swiss Institute for Art Research, 1994), p. 86. See also Lionel Gossman, "Basle, Bachofen, and the Critique of Modernity in the Second Half of the Nineteenth Century," in *Journal of the Warburg and Courtauld Institutes* (Volume 47, 1984), p. 141; and Robert Lougee, *Paul De Lagarde 1827–1891: A Study in Radical Conservatism in Germany* (Cambridge: Harvard University Press, 1962), pp. 227–230.
6. *MDR*, p. 236.
7. See Theodore Ziolkowski, *The View from the Tower* (Princeton: Princeton University Press, 1998).
8. For more, see M.H. Kölbing, "Wie Karl Gustav Jung Basler Professor Würde," *Basler Nachrichten* (September 26, 1954, p. 39); see also Aniela Jaffe. "Details about C.G. Jung's Family," *Spring 1984*, pp. 35–43 and Albert Oeri, "Some Youthful Memories," in *C.G. Jung Speaking: Inteviews and Encounters* [hereafter *CGJS*], ed. William McGuire (Princeton: Princeton University Press), pp. 3–10.
9. In Frederick Gregory, *Nature Lost: Natural Science and the German Theological Traditions of the Nineteenth Century* (Cambridge: Harvard University Press, 1992), p. 37.
10. Foreward to von Koenig-Fachsenfeld, "Wandlungen des Traumproblems von der Romantik bis zur Gegenwart," CW 18, p. 775.
11. *The Zofingia Lectures* (Princeton: Princeton University Press, 1983), p. 96.
12. Ibid., p. 99.
13. Ibid., p. 111.

14. See Zofingia Report 1821–1902 (Basel: Buchdruckerei Kreis, 1902).
15. Letter to Henry Corbin, *Letters, Volume II* (Princeton: Princeton University Press, 1973), p. 115; emphasis in the original.
16. *Letters, Volume I* (Princeton: Princeton University Press, 1973), p. 377.
17. In William McGuire, *Bollingen: An Adventure in Collecting the Past* (Princeton: Princeton University Press, 1982), pp. 23–24.
18. Robert F. Davidson, "Rudolf Otto's Interpretation of Religion," *The Review of Religion* (November 1940), p. 55. Davidson elaborated this into a book by the same name published by Princeton University Press in 1947.
19. Lionel Gossman, *Basel in the Age of Burckhardt* (Chicago: University of Chicago Press, 2000), p. 117. See also Thomas Albert Howard, *Religion and the Rise of Historicism* (Cambridge: Cambridge University Press, 2000), especially pp. 51–70.
20. Gregory, *Nature Lost*, pp. 75–76.
21. Quoted in ibid., p. 79.
22. *The Zofingia Lectures*, p. 110.
23. Ibid., pp. 108–109.
24. Ibid., p. 97.
25. *Letters, Volume II*, pp. 87–91.
26. Marilyn Nagy, *Philosophical Issues in the Psychology of C.G. Jung* (Albany: State University of New York Press, 1991).
27. Paul Means, *Things that are Caesar's* (New York: Round Table Press, 1935), p. 35. Rudolf Otto capitulated to this trend and the influence of his friend Jacob Hauer when he published *Gottheit und Gottheiten der Arier* (Giessan: A. Topelmann, 1932).
28. *The Zofingia Lectures*, p. 44. This youthful distress about Germany's moral decline resurfaced in 1934 when Jung wrote a book introduction. "Schleich thus paid tribute to the scientific past and to the spirit of the Wilhelmine era, when the authority of science swelled into blind presumption and the intellect turned into a ravening beast" (CW 18, p. 466).
29. *The Zofingia Lectures*, p. 95.
30. Ibid., p. 44.
31. Ibid., p. 65.
32. *CGJS*, p. 93. In this he was similar to De Wette "who was completely in tune with the ideology of the Basel governments of the time." Gossman, *Basel in the Age of Burckhardt*, footnote 13, p. 471.
33. *MDR*, p. 73.
34. CW 1, p. 17.
35. See Gossman, *Basel in the Age of Burckhardt*, pp. 55–58. Schleiermacher's upbringing in a Pietist household was a formative influence on his later views on religion, see Howard, *Religion and the Rise of Historicism*, pp. 54–55.
36. See Paul Jenkins, "CMS' Early Experiment in Inter-European Cooperation," in *Church Missionary Society and the World Church 1799–1999* (Basel: Basel Mission, 1999).
37. *The Zofingia Lectures*, pp. 24–25. Strauss came from the same town in Württemberg as Kerner. See also Henri Ellenberger, *The Discovery of the Unconscious* (New York: Basic Books, 1970), pp. 78–81.

38. *The Zofingia Lectures*, p. 37. Jung critiqued Ritschl's appropriation of a pietistic understanding of Christ in his January 1899 lecture (pp. 99–102).
39. See F.X. Charet, *Spiritualism and the Foundation of C.G. Jung's Psychology* (Albany: State University of New York Press, 1993); James Hillman, "Some Early Background to Jung's Ideas: Notes on *C.G. Jung's Medium* by Stephanie Zumstein-Preiswerk," in *Spring 1976*, pp. 123–136; and "C.G. Jung and the Story of Helene Preiswerk: A Critical Study with New Documents," in *Beyond the Unconscious*, ed. Mark Micale (Princeton: Princeton University Press, 1993), pp. 291–305.
40. *MDR*, p. 91.
41. Ibid., p. 40.
42. Ibid., p. 91.
43. Ibid., p. 25.
44. Ibid., p. 104f; and Aniela Jaffe, ed., *Word and Image* (Princeton: Princeton University Press, 1979), p. 37.
45. *MDR*, pp. 104–106 and *Letters, Volume I*, pp. 180–182.
46. Ronald Hayman, *A Life of Jung* (London: Bloomsbury Press, 1999), p. 47.
47. Photos in Jaffe, *Word and Image*, pp. 13 and 33.
48. *MDR*, p. 52.
49. Ibid., p. 50.
50. Ibid., pp. 313–314.
51. Ibid., pp. 11–13.
52. Ibid., p. 11.
53. *MDR*, p. 288.
54. See Ernest Jones, *Sigmund Freud: Four Centenary Addresses* (New York: Basic Books, 1956), p. 109.
55. Gossman, *Basel in the Age of Burckhardt*, p. 73. For the German background, see Fritz Ringer, *The Decline of the German Mandarins* (Hanover, NH: University Press of New England, 1990), pp. 14–42.
56. *CGJS*, p. 207.
57. *MDR*, p. 32.
58. Another precursor of Jung's anti-materialistic critique of science was Goethe's English contemporary William Blake. In their effort to banish superstition Enlightenment thinkers belittled the human faculties of feeling and imagination. In response Blake wrote:

    The Atoms of Democritus
    And Newton's Particles of light
    Are sands upon the Red sea shore,
    Where Israel's tents do shine so bright.

    [*Portable Blake* (New York: Penguin, 1976), p. 142]
59. *CGJS*, p. 209.
60. *MDR*, p. 101.
61. See Timothy Lenoir, *The Strategy of Life* (Chicago: University of Chicago Press, 1982), pp. 172–194; and Adolf Portmann, "Jung's Biology Professor: Some Recollections," in *Spring 1976*, pp. 148–154.
62. *MDR*, pp. 194–195.
63. See Schleich's autobiography *Those Were Good Days!* (London: George Allen and Unwin, 1935), p. 182. Jung may have been introduced to his work by Oscar

Schmitz who dedicated *Die Weltanschauung des Halbgebildeten* (Munich: Georg Mueller, 1914) to Schleich (1859–1922).
64. Ibid., p. 36.
65. Ibid., p. 35.
66. *Transcendental Physics* (1881) has been reprinted by the Kessinger Publishing Company in Montana.
67. The Congress also figured in the séances with his cousin Helly that Jung was attending. She claimed that the spirit of her grandfather Samuel Preiswerk had told her to convert the Jews to Christianity and lead them to Palestine (Hayman, *The Life of Jung*, p. 42). Jung's anti-Semitic sentiment was in contrast to the philo-Semitism found in both branches of his family that was derived from their devotion to biblical philology.
68. See Hans Liebeschütz, "Das Judentum im Geschichtsbild Jacob Burckhardts," in the *Yearbook of the Leo Baeck Institute IV* (1959), especially pp. 73–80; and Aram Mattioli, "Jacob Burckhardts Antisemitismus. Eine Neuinterpretation aus mentalitätsgeschichtlicher Sicht," in *Schweizerische Zeitschrift für Geschichte* (Volume 49, 1999), pp. 496–529. To show the extent of social restrictions placed upon Jews Jung mentioned in his Zarathustra Seminar that until 1865 Jews had to produce a yellow identity card if they wanted to enter the city of Basel. (*Zarathustra Seminar*, pp. 548–549).
69. Gossman, *Basel in the Age of Burckhardt*, p. 244.
70. Harry Kessler, *Berlin in Lights: The Diaries of Count Harry Kessler, 1918–37* (New York: Grove Press, 1999), p. 324.
71. Gossman, *Basel in the Age of Burckhardt*, p. 238.
72. Ira Progoff, *Jung's Psychology and Its Social Meaning* (New York: Grove Press, 1953), p. 34. In his address at Jung's memorial service Hans Schär said "[Burckhardt's] broad humanistic outlook found its continuation in Jung's work under new headings" (Analytical Psychology Club of London, privately printed, 1961), p. 19.
73. *The Greeks and Greek Civilization* (New York: St. Martin's Press, 1998) and *Reflections on History* (Indianapolis: Liberty Classics, 1979).
74. Burckhardt, *Reflections on History*, p. 37. For Jung's use of the phrase, see CW 13, p. 108, CW 15, p. 55, CW 16, p. 124, and CW 17, pp. 88–89.
75. C.G. Jung, *Zarathustra Seminar* (Princeton: Princeton University Press, 1988), p. 274.
76. Friedrich Nietzsche, *The Portable Nietzsche* (New York: Viking Press, 1970), p. 685.
77. Jung, *Zarathustra Seminar*, p. 1301.
78. *Erinnerungen, Träume, Gedanken* (Olten Freiburg im Breisgau: Walter Verlag, 1987), pp. 102–104.
79. Ibid., p. 635.
80. CW 1, pp. 82–84.
81. *MDR*, p. 102.
82. Ibid., p. 103.
83. Gustav Steiner, "Erinnerungen an Carl Gustav Jung aus der Studentenzeit," in the *Basler Stadtbuch 1965* (Basel: Verlag Helbing & Lichtenhahn), p. 157.
84. Jung, *Zarathustra Seminar*, pp. 1191–1192. See also Joel Ryce-Menuhin, "The Symbolists in Art and Literature: their Relationship to Jung's Analytical Psychology," in *Harvest* (Volume 41, No. 1 [1995]), pp. 54–62.

85. CW 6, pp. 166–272. Spitteler received the Nobel Prize for literature in 1919 when a neutral Swiss was a particularly diplomatic choice in the aftermath of World War I. Jung acknowledged that Spitteler resisted a psychological interpretation of his work, maintaining "that his *Olympian Spring* meant nothing, and that he could just as well have sung 'May is come, tra-la-la-la-la'" (CW 15, p. 94).
86. Quoted in Ralph Freeman, *Herman Hesse: Pilgrim of Crisis* (New York: Pantheon Books, 1978), p. 89.
87. Gary Stark, *Entrepreneurs of Ideology: Neoconservative Publishers in Germany, 1890–1933* (Chapel Hill: The University of North Carolina Press, 1981), p. 79.
88. *MDR*, p. 16.
89. Ibid., p. 29. The likely picture is one of an ibex by Johann Tischbein (1751–1829) whose large-scale drawings of animal heads were widely reproduced. See Plate 44 in *In the Footsteps of Goethe* (Düsseldorf: C.G Bornier, 1999), p. 158.
90. Carl Gustav Carus, *Nine Letters on Landscape Painting* (Los Angeles: Getty Research Institute, 2002), p. 91.
91. Jaffe, *Word and Image*, p. 43. For an analysis of another of Jung's landscapes, see *Analytische Psychologie* 18 (1987): "Ein unveroffentlichtes Bild von C.G. Jung: 'Landschaft mit Nebelmeer,'" by Katrin Luchsinger (pp. 298–302); and "Betrachtung eines Bildes von C.G. Jung," by Ursula Baumgart (pp. 303–312).
92. See Robert Rosenblum, Maryanne Stevens, and Ann Dumas, *1900: Art at the Crossroads* (New York: Solomon Guggenheim Museum, 2000).
93. Quoted in James Hillman "Some Early Background to Jung's Ideas," in *Spring 1976*, p. 130.
94. Arthur Gold and Robert Fitzdale, *The Divine Sarah* (New York: Knopf, 1991), pp. 294–295. Freud, also a Bernhardt fan, attended a performance of *Theodora* when he was in Paris in 1885 and kept a photograph of her in his office (p. 4).
95. Cornelia Otis Skinner, *Madame Sarah* (Boston: Hougton Mifflin, 1967), p. 231.
96. Jung Papers in Library of Congress (I: 66, p. 33; author's translation). In the *Visions Seminar* he wrote, "But at the same time, mind you, that this degeneration of pagan culture was taking place, a new style was coming up with its own particular beauty and proportion, the art of the early Christians, the Byzantine art of Ravenna" (p. 1349).
97. See Daniel Gasman, *The Scientific Origins of National Socialism* (Mac Donald [London] and American Elsevier [New York], 1971), pp. 71–76, for the conservative artistic agenda of Ernst Haeckel and the Monist League.

2 Freud and the War Years

1. *MDR*, p. 107. Von Müller (1858–1941) was a professor at Marburg before going to Basel in 1899. He did pioneering studies of metabolism and was the author of many books. In 1922 he was made a member of the Kaiserlich Leopold-Carolinisch Deutsche Akademie der Naturforscher.
2. Quoted in Linda Donn, *Freud and Jung: Years of Friendship, Years of Loss* (New York: Scribner's & Sons, 1988), p. 77.
3. *The Red Book* (New York: W.W. Norton, 2009).

4. See Lynn Gamwell and Richard Ellis, eds., *Sigmund Freud and Art* (State University of New York and the Freud Museum London, 1989), especially Donald Kuspit, "A Mighty Metaphor: The Analogy of Archeology and Psychoanalysis," pp. 133–151. Also, Suzanne Cassirer Bernfeld, "Freud and Archeology," in *American Imago* (Vol. 8:2, 1951), pp. 107–128.
5. C.G. Jung, *Analytical Psychology: Notes on the Seminar Given in 1925* (Princeton: Princeton University Press, 1989) [hereafter *APS*], p. 22.
6. *MDR*, pp. 158–159.
7. Ibid., p. 161; emphasis in the original.
8. Ibid., pp. 163–164.
9. See *Treasures from Basel* (Maastricht, The Netherlands: The European Fine Art Foundation, 1995).
10. *MDR*, pp. 160–161.
11. *Freud/Jung Letters* (hereafter, *F/J L's*) (Princeton: Princeton University Press, 1974), p. 258.
12. C.G. Jung, *Psychology of the Unconscious* (New York: Dodd, Mead and Co., 1946), p. 3.
13. *MDR*, p. 172.
14. Letter quoted in CW 5, p. 32, and in *The Letters of Jacob Burckhardt*, edited and translated by Alexander Dru (London: Routledge & Kegan Paul, 1955), p. 116.
15. *F/J L.'s*, p. 279; emphasis in the original.
16. Ibid., p. 269.
17. Ibid., p. 264.
18. Ibid., p. 439.
19. CW 17, p. 154.
20. Sigmund Freud, *The History of the Psychoanalytic Movement* (New York: Norton, 1966), p. 43.
21. *A Psycho-Analytical Dialogue: The Letters of Sigmund Freud and Karl Abraham, 1907–1926* (New York: Basic Books, 1965), p. 34; see also John Kerr, *A Most Dangerous Method* (New York: Knopf, 1993), pp. 383–387.
22. *The Complete Correspondence of Sigmund Freud and Ernest Jones, 1908–1939*, R. Andrew Paskauskas ed., (Belknap Press, 1993), p. 266. Freud later flipped the ethnic alliances when he wrote to Jones about a "Welsh-Jewish anti-Aryan alliance" (ibid., p. 266).
23. Aniela Jaffe, *Word and Image*, p. 47.
24. Saul Rosenzweig, *The Historic Expedition to America (1909): Freud, Jung, and Hall the Kingmaker* (St. Louis: Rana House, 1994), p. 63. Another is found in Ernest Jones, *The Life and Work of Sigmund Freud, Volume Two* (New York: Basic Books, 1955), p. 102. For a criticism, see Paul Roazen, *Freud and His Followers* (New York: Knopf, 1976), p. 262. The level of sarcasm that characterized the early years of psychoanalysis is evident in the remark by Fritz Wittels that Freud's friendship with Jung revealed his "fondness for bullet heads" in his *Sigmund Freud: His Personality, His Teaching, and His Friendship* (New York: Dodd, Mead, & Co., 1924), p. 177.
25. *Ernest Jones, The Life and Work of Sigmund Freud, Volume Two* (New York: Basic Books, 1953), p. 163.
26. Wittels, *Sigmund Freud: His Personality, His Teachings, and His Friendship*, p. 13.
27. Vincent Brome, *Ernest Jones: Freud's Alter Ego* (New York: Norton, 1983), p. 55.
28. *Freud/Abraham L.'s*, p. 46.

29. F/J L.'s, p. 306.
30. Ibid., "depreciatory Viennese criterion" (p. 526) and "Eitingon's vapid intellectualism" (p. 262).
31. Jung, *Psychology of the Unconscious*, p. 96 [deleted from later editions].
32. *F/J L.'s*, pp. 215, 293; *Freud/Abraham Letters*, p. 139.
33. *A Psycho-Analytic Dialogue*, p. 46; *F/J L.'s*, p. 215.
34. Jung, *Psychology of the Unconscious*, pp. 40–41.
35. See chapter IV, "Monism, The Corporate State, and Eugenics," in *The Scientific Origins of National Socialism* by Daniel Gasman (New York: American Elsevier, 1971). Also, chapter 6, "The Science of Race," in *Toward the Final Solution* by George Mosse (New York: Howard Fertig, 1978); and Sheila Weiss, *Race Hygiene and National Efficiency: The Eugenics of Wilhelm Schallmayer* (Berkeley: University of California Press, 1987).
36. Sander Gilman, *Freud, Race, and Gender* (Princeton: Princeton University Press, 1993), pp. 32–33.
37. CW 6, p. 508.
38. Quoted by Mortimer Ostow, "Letter to the Editor," in *International Review of Psychoanalysis* (Volume 4, 1977), p. 377. See also Renate Schäfer, "Zur Geschichte des Wortes 'zersetzen,'" in *Zeitschrift für Deutsche Wortforschung* (Volume 18, 1962), pp. 40–80.
39. Ludwig Binswanger, *Sigmund Freud: Reminiscences of a Friendship* (New York and London: Grune & Stratton, 1957), p. 46.
40. Maeder to Freud, October 24, 1912 (Freud Collection, Library of Congress).
41. *The Correspondence of Sigmund Freud and Sandor Ferenczi, Volume One 1908–1914*, Eva Brabant, Ernst Falzeder, Patrizia Giampieri-Deutsch, ed. and trans., (Cambridge: Belknap Press, 1993), pp. 490–491.
42. *F/J L.'s*, p. 550.
43. Quoted in Richard Noll, *The Aryan Christ* (New York: Random House, 1997), p. 114.
44. See Jung, *Psychology of the Unconscious*, pp. 153–154 (emphasis in the original); and Hans Walser, "An Early Psychoanalytic Tragedy—J.J. Honegger and the Beginnings of Training Analysis," in *Spring 1974*, pp. 243–255.
45. CW 18, p. 37.
46. John Beebe and Ernst Falzeder, ed., *The Jung-Schmid Letters* (forthcoming, Philemon Series).
47. Quoted in Louis Shaeffer, *O'Neil: Son and Artist* (Boston: Little, Brown, & Co., 1973), p. 245. Jack London was also deeply affected by the book; he told his wife that he was "standing on the edge of a world so new, so wonderful that I am almost afraid to look over into it." Quoted in John Boe, "Jack London, The Wolf, and Jung," in *Psychological Perspectives* (Volume 11, No. 2; Fall 1980), p. 133. See also Alex Kershaw, *Jack London: A Life* (New York: St. Martin's Press, 1997), pp. 286–289. For the New York scene, see Nathan Hale, *The Rise and Crisis of Psychoanalysis in the U.S.: Freud and the Americans 1917–85* (New York: Oxford University Press, 1992), pp. 60–62 and 68–69; and William Scott and Peter Rutkoff, *New York Modern* (Baltimore: Johns Hopkins University Press, 1999), pp. 73–81.
48. See Edward Foote, "Who was Mary Foote?" in *Spring 1974*, especially pp. 258–262; and Dana Sue McDermott, "Creativity in the Theatre: Robert Edmond Jones and C.G. Jung," *Theatre Journal* (May 1984), pp. 213–230.

49. CW 4, p. 292.
50. Ellenberger, *The Discovery of the Unconscious*, pp. 810–814. See CW 18, pp. 427–429, for the texts. For more on Jung's relationship to this publisher, see Paul Bishop, "On the History of Analytical Psychology: C.G. Jung and the Rascher Verlag: Part I and Part II," in *Seminar* 34:3 (September 1998), pp. 256–279 and 34:4 (November 1998), pp. 364–387.
51. Susan Hirsch, "Hodler as Genevois, Hodler as Swiss," in *Ferdinand Hodler, Views and Visions* (Zurich: Swiss Institute for Art Research, 1994), p. 68.
52. Alphons Maeder, *Ferdinand Hodler: Eine Skizze* (Zurich: Rascher Verlag, 1916).
53. Jung lectured to the Psychology Club Zurich on "Aurelia" in 1945; see CW 18, p. 779. *She* made the made most lasting impact on him; the best edition is *The Annotated She* with introduction and notes by Norman Etherington (Bloomington: Indiana University Press, 1991). What has escaped notice is that the itineraries of Jung's two trips to Africa so closely follow those in *Atlantide* (North Africa) and *She* (East Africa) that it seems likely that he planned them, consciously or unconsciously, with the novels in mind.
54. CW 18, p. 762.
55. See Mike Mitchell, *Vivo: The Life of Gustav Meyrink* (Sawtry, U.K.: Dedalus, 2008). For more on *The Golem*, see *Golem: Danger, Deliverance, and Art*, Emily Bilski, ed. (New York: The Jewish Museum, 1988), especially pp. 36 and 56–60. For Wolff's opinion of Jung's literary taste, see *Kurt Wolff: A Portrait in Essays and Letters* (Chicago: University of Chicago Press, 1991), p. 13. For more on Jung's distinction between "visionary" and "psychological" modes of artistic creation, see Morris Philipson, *Outline of a Jungian Aesthetics* (Evanston: Northwestern University Press, 1963), pp. 103–131.
56. Edwin Slossen, *Major Prophets of Today* (Boston: Little, Brown, & Co., 1914), p. 7.
57. Jung, *Psychology of the Unconscious*, p. 64 (CW 5, p. 50). For more on Maeterlinck, see Frantisek Deak, *Symbolist Theater: The Formation of an Avant-Garde* (Baltimore: Johns Hopkins University Press, 1993), especially pp. 158–167.
58. *Jung Letters, Vol. I*, p. 286; CW 6, p. 191.
59. CW 10, p. 27.
60. Jones, *The Life and Work of Sigmund Freud, Volume Two*, p. 39.
61. See Stark, *Entrepreneurs of Ideology*, pp. 14–19 and 58–110. See also Meike Werner, "Provincial Modernism: Jena as a Publishing Program," in *Germanic Review* (Volume 76, No. 4, Fall 2001), pp. 319–334.
62. Jung, *Psychology of the Unconscious*, p. 490 (f.n. 1) and p. 552 (f.n. 112).
63. For more on Gross, the Cosmics, and Schwabing, consult three books by Martin Green, *The von Richthofen Sisters* (New York: Basic Books, 1974), pp. 32–100; *Mountain of Truth: The Counter Culture Begins, Ascona 1900–1920* (Hanover/ London: University Press of New England, 1986); and *Otto Gross, Freudian Psychoanalyst 1877–1920* (Lewiston, NY: The Edwin Mellen Press, 1999). See also, David Clay Large, *Where Ghosts Walked: Munich's Road to the Third Reich* (New York: W.W. Norton and Co., 1997), pp. 3–34.
64. *F/J L.'s*, p. 153.
65. Ibid., p. 156.

66. Quoted in Gottfried Heuer, "Jung's Twin Brother. Otto Gross and Carl Gustav Jung," *Journal of Analytical Psychology* (Volume 46, No. 3, 2001), p. 670.
67. Aldo Carotenuto, *A Secret Symmetry* (New York: Pantheon Books, 1982), p. 107.
68. *F/J L.'s*, p. 207.
69. Quoted in Nina Kindler, "G.R. Heyer in Deutschland," in *Die Psychologie des 20. Jahrhunderts, Band III: Freud und die Folgen,* ed. Dieter Eicke (Zurich: Kindler Verlag, 1977), p. 82.
70. See Stark, *Entrepreneurs of Ideology*, pp. 19–22 and 120–124. Lehmanns Verlag, which published medical books by Friedrich von Müller, was disbanded after World War II as a Nazi-affiliated press; see Robert Proctor, *The Nazi War on Cancer* (Princeton: Princeton University Press, 1999), f.n. 50, p. 285. Jung reviewed Heyer's book for the *Europäische Revue* (IX: 10, October 1933).
71. See Schmitz's novel *Bürgerliche Boheme*, ed. by Monika Dimpfl and Carl-Ludwig Reichert (Bonn: Weidle Verlag, 1998) and the second volume of his memoirs *Daemon Welt* (Munich: Georg Müller, 1926). Also, Georg Fuchs, *Sturm und Drang in München* (Munich: Verlag Callwey, 1936), pp. 91–96. Just how complicated family relations were in Germany at the time is shown by the fact that her brother Count Ernst zu Reventlow was an early supporter of the Nazis and helped Jacob Wilhelm Hauer start the German Faith Movement.
72. *MDR*, pp. 110–111. For the art scene, see Maria Makela, *The Munich Secession: Art and Artists in Turn-of-the-Century Munich* (Princeton: Princeton University Press, 1990); see also Peg Weiss' *Kandinsky in Munich: The Formative Jugendstil Years* (Princeton: Princeton University Press, 1979); and *Kandinsky in Munich: 1896–1914* (New York: The Solomon Guggenheim Museum, 1982).
73. *F/J L.'s*, p. 243.
74. Ibid., p. 386.
75. Ibid., p. 478.
76. Wassily Kandinsky, *Concerning the Spiritual in Art* (New York: Dover Publications, 1977), p. 2.
77. See Suzanne Marchand, *German Orientalism in the Age of Empire: Religion, Race and Scholarship* (New York: Cambridge University Press, 2009); and *Down from Olympus: Archaeology and Philhellenism in Germany, 1750–1970* (Princeton: Princeton University Press, 1996).
78. *MDR*, p. 284. These blue mosaics triggered the hallucination that Jung experienced on his return visit twenty years later.
79. Quoted in Frank Whitford, *Gustav Klimt* (New York: Crescent Books, 1994), p. 86.
80. The most frequently reproduced version is that in *The Red Book*, p. 154. The version that depicts the original dream of Philemon is in Gerhard Wehr, *An Illustrated Biography of C.G. Jung* (Boston and Shaftesbury: Shambhala, 1989), p. 72; the final version is the one he painted at Bollingen and is in Carl Jung, *Man and His Symbols* (Garden City: Doubleday and Co., 1964), p. 198. For further pictorial elaboration, see my "A Pictorial Guide to *The Red Book*," ARAS Connections: Image and Archetype (2010, Issue 1) <info@aras.org>.
81. *MDR*, p. 182. One possible source for the name "Philemon" that has not previously been considered is the Victorian romance *Hypatia* (1853) by Charles Kingsley.

The protagonist of this novel about the late-Roman Neo-Platonic philosopher is a monk named Philammon and in the *Red Book* Jung dialogues with an anchorite named Ammonius. Jung owned a 1902 edition of the book. See also James Heisig, "*The VII Sermones*: Play and Theory," *Spring 1972*, pp. 206–218.

82. Jaffe, *Word and Image*, pp. 69, 70, 73, and 75.
83. See Ronald Hayman, *A Life of Jung* (London: Bloomsbury, 1999), p. 274.
84. *MDR*, pp. 185–187. The identification of the woman as Maria Moltzer was made by Sonu Shamdasani in "Memories, Dreams, Omissions," in *Spring 95*, pp. 127–129. This is based on documents in the Jung Papers, Library of Congress where the individual is identified by the initials "M.M."
85. *MDR*, pp. 181–182 and *APS*, pp. 88–90, 92–98.
86. Kerr, *A Most Dangerous Method*, pp. 468–469.
87. See Fuchs, *Sturm und Drang in München*, pp. 242–243. See also Bram Dijkstra, *Idols of Perversity* (Oxford: Oxford University Press, 1986), pp. 379–401; and Shearer West, *Fin de Siècle: Art and Society in an Age of Uncertainty* (Woodstock: The Overlook Press, 1994), p. 93.
88. See Toni Bentley, *Sisters of Salome* (New Haven: Yale University Press, 2003); Rhonda Gerlick, *Electric Salome: Loie Fuller's Performance of Modernism* (Princeton: Princeton University Press, 2007); and Holly Edwards et al., *Noble Dreams, Wicked Pleasures: Orientalism in America 1870–1930* (Princeton: Princeton University Press/Clark Art Institute, 2000), pp. 207–209 for what the apache hall and its entertainment would have looked like.
89. *F/J L.'s*, p. 229.
90. Hayman, *The Life of Jung*, p. 66. See also James Rice, "Russian Stereotypes in the Freud-Jung Correspondence," *Slavic Review* (Volume 41, Spring 1982), pp. 19–34. It should be remembered that Ivenes, Helly Preiswerk's sub-personality, was a "black-haired woman of markedly Jewish type, clothed in white garments, her head wrapped in a turban" (CW 1, p. 33). This type is represented by the Sephardic beauty drawn by Ephraim Moses Lillien and reproduced in John Efron, *Defenders of the Race: Jewish Doctors & Race Science in Fin-de-Siècle Europe* (New Haven: Yale University Press, 1994), after p. 90. For more on Lillien, see Michael Stanislawski, *Zionism and the Fin de Siècle, Cosmopolitanism and Nationalism from Nordau to Jabotinsky* (Berkeley: University of California Press, 2001), pp. 98–115. Later, in the Zarathustra Seminar Jung announced that "Every true Protestant has a Jewish anima who preaches the Old Testament" (p. 922).
91. The article first appeared in *Die Psychologie* (Bern), Heft 7/8, Band III, 1951 and then was privately printed in English translation for the Student Association of the C.G. Jung Institute, Zurich, 1956.
92. Walter Laqueur, *Young Germany: A History of the German Youth Movement* (New Brunswick and London: Transaction Books, 1984), p. 51; emphasis in the original.
93. Green, *Mountain of Truth*, p. 173.
94. *F/J L.'s*, p. 289.
95. C.G. Jung, *Notes of a Seminar given in 1925*, William McGuire ed., (Princeton: Princeton University Press, 1989), [hereafter, *APS*], p. 93.
96. *MDR*, p. 286.
97. *APS*, p. 80.
98. Jung, *Psychology of the Unconscious*, p. 411.

99. *F/J L.'s*, p. 294.
100. "The Letters of C.G. Jung to Sabina Spielrein," *Journal of Analytical Psychology* (Volume 46, No. 1, 2001), p. 190.
101. Ibid., p. 194.
102. *Jung Letters Volume I*, pp. 31–32.
103. *Visions Seminar, Volume II* (Princeton: Princeton University Press, 1997), p. 974; emphasis in the original.
104. In Emil Ermatinger, ed., *Philosophie der Literaturwissenschaft* (Berlin: Junker und Dunnhaupt, 1930), pp. 315–330, now in CW 15, pp. 84–105.
105. Hirsch, "Hodler as Genevois, Hodler as Swiss," pp. 97–98; See also Gustave Le Bon, *Psychology of the Great War* (New Brunswick and London: Transaction Books, 1999), pp. 405–406.
106. *Zarathustra Seminar*, p. 813 and Forrest Robinson, *Love's Story Told: A Life of Henry Murray* (Cambridge: Harvard University Press, 1992), p. 124. (That Jung continued to support an aggressive German foreign policy can be seen in his advocacy for a German invasion of the U.S.S.R. in 1939. [*CGJS*, p. 132]). The vast majority of German intellectuals supported the country's war aims. Shortly after it started ninety-three leading figures in the arts and sciences including Ernst Haeckel, Adolf von Harnack, Max Lieberman, and Max Planck signed a manifesto that defended the invasion of Belgium and the destruction of Louvain. See John Horne and Alan Kramer, *German Atrocities 1914: A History of Denial* (New Haven and London: Yale University Press, 2001), pp. 277–290. For more on the conduct of German intellectuals during the war, see Fritz Ringer, *The Decline of the German Mandarins* (Hanover, NH: University Press of New England, 1990), pp. 180–199; and Klaus Schwabe, "Zur politischen Haltung der Deutschen Professoren im Ersten Weltkreig," *Historische Zeitschrift* (Volume 193, 1961), pp. 601–634.
107. "The Letters of C.G. Jung to Sabina Spielrein," op. cit., p. 188. In the same vein, in a passage deleted from *Psychology of the Unconscious* Jung noted that "one should never forget the harsh speech of the first Napoleon, that the good God is always on the side of the heaviest artillery" (pp. 261–262).
108. *MDR*, p. 176.
109. *Zarathustra Seminar*, pp. 697–698. Jung's unconscious involvement with Germany's leader was repeated before World War II when he dreamed of Hitler (*CGJS*, pp. 180–181).
110. *MDR*, p. 195.
111. *The Red Book*, p. 125.
112. George Mosse, *The Origins of German Ideology* (New York: Grosset and Dunlop, 1964), chapter 7; Lèon Poliakov, *The Aryan Myth* (New York: Barnes and Noble Books, 1996), chapters 10, 11.
113. CW 10, p. 14; emphasis in the original.
114. "The Letters of C.G. Jung to Sabina Spielrein," op. cit., p. 194.
115. Magnus Ljunggren, *The Russian Mephisto: A Study of the Life and Work of Emilii Medtner* (Stockholm: Acta Universitatis Stockholmiensis, Stockholm Studies in Russian Literature, 1994), p. 120.
116. CW 7, p. 175 and footnote 4, p. 147; also, *APS*, pp. 33 and 37 and CW 9i, p. 124. For more on how race was treated by French intellectuals, see Christopher

E. Forth, *The Dreyfus Affair and the Crisis of French Manhood* (Baltimore and London: The Johns Hopkins University Press, 2004).

117. CW 6, p. 487.
118. Keyserling file, p. 18, C.G. Jung Oral History Archive, Countway Medical Library, Harvard University, Boston. Also, Blüher said that "Schmitz showed himself politically in every situation to be an out-and-out Prussian war fanatic" (in *Werke und Tage* [München: Paul List Verlag, 1953], p. 361).

3 Jung's Post-Freudian Network

1. See Beryl Pogson, *Maurice Nicoll: A Portrait* (New York: Fourth Way Books, 1987); Ian Begg, "Jung's Lost Lieutenant," in *Harvest* (Volume 22, 1976), pp. 78–90; and William McGuire, "Firm Affinities," in *Journal of Analytical Psychology* (Volume 40, No. 3, 1995), pp. 301–326.
2. Barbara Hannah, *Jung: His Life and Work* (New York: Putnam and Sons, 1976), pp. 150–153. This summary is based on notes in the Kristine Mann Library.
3. CW 7, pp. 19–40.
4. Hermann Keyserling, *Europe* (New York: Harcourt, Brace, 1928), p. 333.
5. William McDougall, *Is America Safe for Democracy?* (New York: Charles Scribner's and Sons, 1921), pp. 125–27.
6. Pogson, *Maurice Nicoll*, p. 49.
7. McDougall, *Is America Safe for Democracy?* p. 126; see also his *Outline of Abnormal Psychology* (New York: Charles Scribner's and Sons, 1926), p. 190.
8. Jung used this term in his 1914 lecture to the British Psychical Society where he possibly met McDougall. See CW 3, p. 140, f.n. 16.
9. See McDougall's *Modern Materialism and Emergent Evolution* (London: Metheun, 1934), p. 217.
10. William McDougall, *Psychoanalysis and Social Psychology* (London: Metheun and Co., 1936), pp. 110–111.
11. Thomas Gosset, *Race* (New York: Oxford University Press, 1997), p. 177. For more on the decline of McDougall's reputation, see Russell Jones, "Psychology, History, and the Press, The Case of William McDougall and the *New York Times*," in *American Psychologist* (Volume 42, Number 10, October 1987), pp. 931–940.
12. *CGJS*, p. 30.
13. *Forum* (Volume 83, No. 4), p. 196.
14. *MDR*, pp. 270–272.
15. *Forum* (Volume 83, No. 4), p. 196. Jung's redating from "1500" to "1200" years seems to have been due to his effort to account for Charlemagne's forced conversion of the Saxons to Christianity.
16. George Santayana, *Egotism in German Philosophy* (New York: Scribner's and Sons, n.d.), pp. 149–150.
17. See John Horne and Alan Kramer, *German Atrocities 1914: A History of Denial* (New Haven: Yale University Press, 2001), pp. 217–221; and Gustave Le Bon, *Psychology of the Great War* (New Brunswick: Transaction, 1999), pp. 388–389.

The group fantasy about the Germans acting like "Huns" did not begin with Allied propagandists during the war but with Kaiser Wilhelm II himself. In his speech to the troops being sent to China to help suppress the Boxer Rebellion (1900) he encouraged them to help Germany find its "place in the sun" by acting as ruthless as Huns.

18. Hans Sluga, *Heidegger's Crisis: Philosophy and Politics in Nazi Germany* (Cambridge: Harvard University Press, 1993), p. 98.
19. Fritz Ringer, *The Decline of the German Mandarins* (Hanover, NH : University Press of New England, 1990), p. 3.
20. *Letters: Volume I*, pp. 39–41.
21. Paul Means, *Things that are Caesar's* (New York: Round Table Press, 1935), p. 57.
22. *MDR*, p. 313.
23. Oscar A.H. Schmitz, *Ergo Sum* (Munich: Georg Müller, 1926), pp. 301–302.
24. CW 10, p. 184; emphasis in the original.
25. Marie Louise von Franz, *Psychotherapy* (Boston: Shambhala, 1993), pp. 310–311.
26. See Walter Laqueur, *Young Germany: A History of the German Youth Movement* (New Brunswick: Transaction, 1984), chapters 11–13.
27. For more, see Walter Struve, *Elites Against Democracy* (Princeton: Princeton University Press, 1973), pp. 299–304, which stresses Keyserling's cosmopolitanism.
28. Richard Noll, *The Aryan Christ* (New York: Random House, 1997), p. 93.
29. See CW 6, p. xv; and *CGJS*, pp. 82–84.
30. See Geoffrey Field, *The Evangelist of Race* (New York: Columbia University Press, 1981).
31. Lèon Poliakov, *The Aryan Myth* (New York: Barnes and Noble, 1996), pp. 275, 285.
32. Keyserling, *Europe*, p. 156.
33. Houston Stuart Chamberlain, *Foundations of the Nineteenth Century* (London: Bodley Head, 1914), p. 539.
34. *Zarathustra Seminar*, p. 814. It is possible that Jung remembered this from his own reading of Chamberlain's work of which he owned a copy.
35. Hermann Keyserling, *The Recovery of Truth* (New York: Harper's Bros. Publishing, 1929), p. 399.
36. Quoted in Struve, *Elites Against Democracy*, p. 312.
37. Harry Kessler, *Berlin in Lights: The Diaries of Count Harry Kessler, 1918–37* (New York: Grove Press, 1999), p. 455.
38. See Joseph Campbell's Introduction to *Myth, Religion, and Mother Right* (Princeton: Princeton University Press, 1973).
39. C.G. Jung, *Contributions to Analytical Psychology* (New York: Harcourt, Brace, 1928), p. 118; and CW 10, par. 53.
40. McDougall, *Psychoanalysis and Social Psychology*, p. 110.
41. Jung, *Contributions to Analytical Psychology*, p. 139.
42. *APS*, p. 107.
43. Roderick Stackelberg, *Idealism Debased: From Völkisch Ideology to National Socialism* (Kent, OH: Kent State University Press, 1981), chapter 12. See also Richard Gray, *About Face: German Physiognomic Thought from Lavater to Auschwitz* (Detroit: Wayne State University Press, 2004).

44. Jung, *Contributions to Analytical Psychology*, p. 139.
45. Margaret Gildea, "Jung As Seen By An Editor, A Student, And A Disciple," in *Carl Jung, Emma Jung, and Toni Wolff* (Analytical Psychology Club of San Francisco, 1982), p. 24.
46. Max Scheler, *Man's Place in Nature* (New York: Noonday Press, 1971), p. 85.
47. Gary Stark, *Entrepreneurs of Ideology: Neoconservative Publishers in Germany, 1890–1933* (Chapel Hill: The University of North Carolina Pres, 1981), pp. 4–6.
48. Ibid., p. 9.
49. CW 7, par. 240.
50. Struve, *Elites Against Democracy*, p. 309. All Noll managed to say in his "scholarly" analysis of the various journals and newspapers in which Jung published was that "These publications were somewhat akin to today's politically conservative *Reader's Digest...*" *The Jung Cult* (Princeton: Princeton University Press, 1994), f.n. 124, p. 365.
51. In "Die Geistige Problem Europa von Heute" (Vienna: Verlag der Wila, 1922).
52. Schmitz, *Ergo Sum*, p. 329.
53. Armin Mohler, *Die Konservative Revolution in Deutschland* (Darmstadt: Wissenschaftliche Buchgesellschaft, 1994), p. 293.
54. *Neue Freie Presse*, Vienna, February 19, 1928.
55. Guido Müller, *Deutsch-Franzosische Gesellschaftsbeziehungen nach dem Ersten Weltkrieg* (Aachen: Habilitationsschrift, 1997), pp. 476–479.
56. CW 10, pp. 75–76.
57. Ibid., p. 77.
58. Ibid., p. 85.
59. Ibid., p. 87.
60. Ibid.
61. Ibid., p. 90.
62. CW 8, p. 34 (par. 652).
63. Richard Cavendish, *Encyclopedia of the Unexplained* (New York: McGraw-Hill Books, 1974), p. 46.
64. Hermann Rauschning, *Hitler Speaks* (London: Thornton, Butterworth, 1940), p. 240.
65. For more, see Sonu Shamdasani's Introduction to *The Psychology of Kundalini Yoga* (Princeton: Princeton University Press, 1996).
66. L. Sprague De Camp, *Lost Continents* (New York: Gnome Press, 1954), p. 86.
67. CW 17, p. 167.
68. Ibid., p. 178.
69. Ibid., p. 185.
70. Ibid., p. 177.
71. CW 18, p. 767.
72. CW 15, pp. 138–139.
73. *Letters: Volume I*, p. 88.
74. Struve, *Elites Against Democracy*, p. 227.
75. Robert Lougee, "German Romanticism and Political Thought," in *The European Past: Vol. 2* (New York: Macmillan, 1964), p. 93.

76. CW 17, p. 171.
77. Anton Kaes, Martin Jay, and Edward Dimenberg, eds., *The Weimar Republic Sourcebook* (Berkeley: University of California Press, 1995), p. 116.
78. *Lexikon zur Parteigeschichte: Vol. 2* (Berlin: Pahl-Pugenstein, 1984), p. 422.
79. Müller, *Deutsch-Franzosische Gesellschaftsbeziehungen*, p. 76.
80. *Lexikon der Konservatismus* (Graz-Stuttgart: Stocker. 1996), p. 162. For more on the von Schnitzlers, see Dietrich Neumann, "The Barcelona Pavilion," in *Barcelona and Modernity* (New Haven: Yale University Press, 2006), pp. 390–399; and *Hell's Cartel: IG Farben and the Making of Hitler's War Machine* (New York: Metropolitan Books, 2008).
81. Leopold Ziegler in *Philosophie der Gegenwart in Selbstdarstellungen: Vol. 4* (1923), p. 208.
82. Edgar Jung, *Die Sinndeutung der deutschen Revolution* (Oldenburg: Stalling, 1933), p. 22.
83. *ZS: Vol. 2*, pp. 1002–1003.
84. Kaes et al., eds., *The Weimar Republic Sourcebook*, pp. 352, 354.
85. Geoffrey Cocks, *Psychotherapy in the Third Reich* (New Brunswick: Transaction Publications, 1997), p. 137.
86. *Letters: Vol. I*, p. 113. Also see Jacobi's reminiscence of a visit paid to Ziegler at his home on Lake Constance by the Jungs, Otto Curtius, and herself that year. (Countway Medical Library, Jung Oral History Archive, Box 3, pp. 18–19.)

4 The Question of Accommodation

1. Hans Sluga, *Heidegger's Crisis: Philosophy and Politics in Nazi Germany* (Cambridge: Harvard University Press, 1993), p. 70.
2. John Burnham and William McGuire, *Jelliffe: American Psychoanalyst and Physician and His Correspondence with Sigmund Freud and C.G. Jung* (Chicago: University of Chicago Press, 1983), p. 249.
3. S. Grossman, "C.G. Jung and National Socialism," *Journal of European Studies* (Volume ix, 1979), p. 233.
4. "Dr. Jung's Farewell, October, 1936," file in the Kristine Mann Library, New York.
5. Geoffrey Cocks, *Psychotherapy in the Third Reich* (2nd edition) (New Brunswick, NJ: Transaction Press, 1997), p. 147.
6. *German Fiction Writers 1885–1913, Part I* [Dictionary of Literary Biography, Volume 66] (Detroit: B.C. Layman Books, 1988), p. 37.
7. Sluga, *Heidegger's Crisis*, pp. 1–4.
8. Ibid., 125–135.
9. Rüdiger Sufranski, *Martin Heidegger, Between Good and Evil* (Cambridge: Harvard University Press, 1998), p. 133.
10. See the November 8 editions of the *Bonner Zeitung*, the *Deutsche Reichs-Zeitung*, and the *General-Anzieger für Bonn*. For more on Rothacker, see Emmanuel Faye, *Heidegger: The Introduction of Nazism into Philosophy* (New Haven & London: Yale University Press, 2009).

11. Ulfried Geuter, *The Professionalization of Psychology in Nazi Germany* (Cambridge: Cambridge University Press, 1992), pp. 42 and 111.
12. Sluga, *Heidegger's Crisis*, p. 96; and Anne Harrington, *Reenchanted Science: Holism in German Culture from Wilhelm II to Hitler* (Princeton: Princeton University Press, 1996), pp. 124f.
13. Jerry Muller, *The Other God that Failed: Hans Freyer and the Deradicalization of German Conservatism* (Princeton: Princeton University Press, 1987), p. 90.
14. See *Transzendenz als Erfarhung, Beitrag und Widerhall: Festschrift zum 70. Geburtstag von Graf Dürckheim* (Weilheim/OBB.: Otto Wilhelm Barth-Verlag, 1966).
15. CW 10, p. 205.
16. *Thomas Mann Diaries 1918–39* (London: Andre Deutsch, 1983), p. 178.
17. Ibid., p. 202.
18. Alan Schom, *Survey of Nazi and Pro-Nazi Groups in Switzerland, 1930–1945* (Simon Wiesenthal Center, 1988), pp. 6, 8.
19. *Neue Zürcher Zeitung*, December 19, 1932.
20. Barbara Hannah, *Jung: His Life and Work* (New York: Putnam and Sons, 1976), p. 211.
21. Paul Means, *Things that are Caesar's* (New York: Round Table Press, 1935), p. 178; and Martin Green, *The von Richthofen Sisters* (New York : Basic Books, 1974), p. 224.
22. Magnus Ljunggren, *The Russian Mephisto: A Study of the Life and Work of Emilii Medtner* (Stockholm: Acta Universitatis Stockholmiensis, Stockholm Studies in Russian Literature, 1994), p. 146.
23. Quoted in Klaus Fischer, *Nazi Germany: A New History* (New York: Continuum, 1995), p. 366.
24. Quoted in Philip Metcalf, *1933* (Sag Harbor, NY: Permanent Press, 1988), p. 122.
25. Michael Fordham, "Memories and Thoughts about C.G. Jung," in *Freud, Jung, Klein: The Fenceless Field* (London: Routledge, 1995), p. 108.
26. CW 18, p. 164.
27. *Deutsches Literatur-Lexikon* (Band I), Wilhelm Kosch ed. (Bern, München: Francke Verlag, 1968), pp. 614–616.
28. See Matthias von der Tann and Arvid Erlenmeyer, eds., *C. G. Jung und der Nationalsozialismus* (Berlin: privately printed by the Deustschen Gesellschaft für Analytische Psychologie, 2nd ed., 1993), pp. 5–7.
29. C.G. Jung, *Two Essays on Analytical Psychology* (New York: Dodd, Mead, and Company, 1928), p. 43.
30. *Letters, Volume I*, pp. 121–122.
31. See Hanspeter Brode, *Benn Chronik* (München: Carl Hanser Verlag, 1978). For more on Benn, see George Mosse, *Germans and Jews* (Detroit: Wayne State University Press, 1987), chapter 6, "Fascism and the Intellectuals."
32. Quoted in *Primal Vision, Selected Writings of Gottfried Benn* (London: Marion Boyars, 1976), pp. xiv–xv.
33. Ibid., pp. 47–48. See also Olga Solovieva, "'Bizarre Epik des Augenblicks': Gottfried Benn's 'Answer to the Literary Emigrants' in the Context of His Early Prose," *German Studies Review* (Volume 33, No. 1, 2010). For more on the legacy

of the African soldiers in the Rhineland, see Michael Burleigh and Wolfgang Wipperman, *The Racial State: Germany 1933–1945* (New York: Cambridge University Press, 1991), pp. 128–130.

34. *Letters, Volume I*, p. 40.
35. *German Fiction Writers 1885–1913*, p. 46.
36. Hermann Rauschning, *Hitler Speaks* (London: Thornton, Butterworth, 1940), pp. 142 and 144.
37. Robert Proctor, *Racial Hygiene: Medicine under the Nazis* (Cambridge: Harvard University Press, 1988), p. 50.
38. Ibid., pp. 178–179.
39. See Geoffrey Cocks, *Psychotherapy in the Third Reich* (New Brunswick: Transaction, 1997), pp. 29–31.
40. *F/J L's*, p. 214, footnote 1.
41. *Freud/Abraham Letters*, p. 382.
42. Gary Stark, *Entrepreneurs of Ideology: Neoconservative Publishers in Germany, 1890–1933* (Chapel Hill: The University of North Carolina Press, 1981), pp. 19–22.
43. James Webb, *The Occult Establishment* (Glasgow: Richard Drew Publishing, 1981), pp. 296–298.
44. Cocks, *Psychotherapy in the Third Reich*, p. 102.
45. Ibid., p. 101.
46. *Letters, Volume I*, p. 132.
47. *CGJS*, p. 39.
48. *Letters, Volume II*, pp. 14–15.
49. *Brockhaus Encyclopedia* (1970).
50. *Letters, Volume I*, p. 135.
51. Cocks, *Psychotherapy in the Third Reich*, p. 134.
52. CW 10, pp. 533–534.
53. Ibid., p. 536.
54. Ibid., pp. 538 and 539.
55. Ibid., p. 537.
56. Ibid.
57. Ibid., p. 540.
58. *Thomas Mann Diaries*, pp. 201 and 235.
59. CW 10, p. 544.
60. See chapter three.
61. *Letters, Volume I*, p. 154.
62. CW 10, p. 541.
63. Quoted in Mortimer Ostow, Letter to the Editor, *International Review of Psycho-Analysis* (Volume 4, 1977).
64. CW 10, p. 538.
65. *Letters, Volume I*, p. 146.
66. Ibid., pp. 160–163.
67. Ibid., pp. 164–165.
68. Ibid., p. 162.
69. James Kirsch, "Reflections at Age Eighty-Four," in *A Modern Jew in Search of a Soul*, ed. Spiegelmann and Jacobson (Phoenix: Falcon Press, 1986), p. 182. See

also Mosse, *Germans and Jews*, chapter 4 ("The Influence of the Volkish Idea on German Jewry").

70. *Analytische Psychologie* (Volume 11, Number 3–4), p. 182. For more background, see Michael Brenner, *The Renaissance of Jewish Culture in Weimar Germany* (New Haven: Yale University Press, 1996).
71. All translations are by Charles Boyd, Berlin.
72. Kirsch, "Reflections at Age Eighty-Four," p. 151.
73. Letter to the author, August 31, 1985.
74. *Letters, Volume I*, p. 163.
75. Jung Oral History Archive (Box 3, p. 64), Countway Medical Library, Boston.
76. William McGuire, *Bollingen: An Adventure in Collecting the Past* (Princeton: Princeton University Press, 1982), pp. 24–26. The author is inaccurate when he states that Hauer "subsequently" founded the German Faith Movement. In fact, Hauer became its führer several weeks before attending the Eranos Conference. Also, McGuire is incorrect when he says that Jung criticized Hauer in his "Wotan" article.
77. Cocks, *Psychotherapy in the Third Reich*, p. 51
78. *Letters, Volume I*, pp. 153–154.
79. William Shirer, *Rise and Fall of the Third Reich* (New York: Simon and Schuster, 1960), p. 215. One can only wonder if attendees at either conference visited the ruins of the Mithraic temple at nearby Heddernheim.
80. C.G. Jung Oral History Archive (Box 17, pp. 16–17), Countway Medical Library, Boston. See also Peter Kamber, *Geschichte zweier Leben—Wladimir Rosenbaum und Aline Valagin* (Zurich: Limmat Verlag, 1990), pp. 168–170.
81. Jung Oral History Archive, Box 17, pp. 10–11, Countway Medical Library, Boston.
82. Cocks, *Psychotherapy in the Third Reich*, p. 134.
83. 1934 *Zentralblatt*, p. 18.
84. *Ziel und Weg* (1934, 12), p. 467.
85. For details, see Benno Müller-Hill, *Murderous Science* (Oxford: Oxford University Press, 1988).
86. *Nova Acta Leopoldina* (Neue Folge Band 2, Heft 5; 1934), pp. 641–643.
87. Carl Ludwig Schleich, *Those Were Good Days!* (London: Allen & Unwin, 1935), p. 163.
88. Ibid., p. 182.
89. *Letters, Volume I*, p. 164. Several weeks later Jung mentioned the review in his Zarathustra seminar of June 27 (p. 138).
90. Proctor, *Racial Hygiene*, pp. 233–234.
91. Ibid., p. 231.
92. CW 18, p. 463.
93. Ibid., p. 465.
94. My translation since the Hull version is inexact.
95. Ibid., p. 466. See also his letter to Gilbert (*Volume I*, p. 164).
96. Walter Struve, *Elites Against Democracy* (Princeton: Princeton University Press, 1973), p. 312.
97. Harry Kessler, *Berlin in Lights: The Diaries of Count Harry Kessler, 1918–37* (New York: Grove Press, 1999), p. 463. See also Paul Valery, *CW 10: History and Politics* [Bollingen Series XLV] (New York: Pantheon Books, 1962), pp. 531–534.

98. CW 10, p. 496.
99. Ibid., p. 499.
100. Carl Ludwig Schleich, *Die Wunder der Seele* (Frankfurt: G.B. Fischer, 1951), p. 299.
101. Jung Oral History Archive (Box 14, pp. 22–23), Countway Medical Library, Boston.

5 Nazi Germany and Abroad

1. *Die Kulturelle Bedeutung Der Komplexen Psychologie* (Berlin: Verlag von Julius Springer, 1935).
2. CW 10, p. 551.
3. See *Letters, Volume I*, pp. 275–276 and 286–288; also, Geoffrey Cocks, *Psychotherapy in the Third Reich* (New Brunswick: Transaction, 1997), pp. 143–144, where van der Hoop is mistakenly identified as the head of the Danish group.
4. See chapter two, footnote 42.
5. *Letters, Volume I*, pp. 204–206.
6. Unpublished film interview conducted on January 24, 1989, in Zurich by Aryeh Maidenbaum, Stephen Martin, and Robert Hinshaw. I want to thank Aryeh Maidenbaum for allowing me to consult it.
7. CW 10, p. 549; emphasis in the original.
8. Ibid., p. 538.
9. Ibid., p. 555.
10. Ibid., p. 552 (par. 1053). Another example occurred in his 1936 Bailey Island Seminar. In response to a question about the influence of racial experience on the development of archetypes Jung said that there was a racial component in the unconscious (*Bailey Island Seminar*, p. 148, privately printed, in the Kristine Mann Library, New York).
11. Matthias von der Tann and Arvid Erlenmeyer, eds., *C. G. Jung und der Nationalsozialismus* (Berlin: Deustschen Gesellschaft für Analytische Psychologie, 1993), p. 24.
12. Heraclitus: pre-Socratic philosopher who's aphoristic, aristocratic philosophy was the subject of Spengler's doctoral dissertation and the source of Jung's concept of *enantiodromia.*
13. The dubious scientific status and criminal misuse of the first of these is well-known; see Benno Müller-Hill, *Murderous Science* (Oxford: Oxford University Press, 1988). For a discussion of Volk psychology, a uniquely German field of study, see Woodruff Smith, *Politics and the Sciences of Culture in Germany 1840–1920* (Oxford: Oxford University Press, 1991), chapter 6.
14. The 1936 *Zentralblatt*, p. 292.
15. The 1937–1938 *Zentralblatt*, p. 301. Another Nazi intellectual who employed Jung's type theory was Max Wundt, son of Wilhelm Wundt, the founder of experimental psychology and colleague of Jacob Hauer at Tübingen University. See his *The Roots of German Philosophy in Clans and Races* (Berlin: Junker and Dunnhaupt, 1944), pp. 34–35.

16. *Letters, Volume I*, p. 238.
17. CW 18, pp. 773–775.
18. The reason for its publication by the company of Julius Springer in Berlin rather than by Rascher, Jung's regular publisher, or another Zurich publisher is unclear.
19. The reviews appeared in the *Europäische Revue* and in the *Zentralblatt*. See CW18, pp. 793–796.
20. *Letters, Volume I*, p. 194.
21. Besides Seifert another student of philosophy who got interested in Jung at this time is J. Meinertz. See the 1937/1938 *Zentralblatt* and *Letters, Volume I*, p. 273. He saw an affinity between Jung's concepts and the existential ontology of Martin Heidegger. He also discussed the work of Leopold Ziegler and referred to his recent book *Traditions*, a book that Jung owned.
22. Heyer's books were published by J.F. Lehmanns Verlag, see Stark, *Entrepreneurs of Ideology*, pp. 19–22. Also, Charles Baudouin, *L'Oeuvre de Jung* (Payot: Paris, 1963), pp. 321–335; and Toni Wolf's "Betrachtung und Besprechung von 'Reich der Seele,'" 1937/1938 *Zentralblatt*, pp. 239–278.
23. I want to thank Matthias van der Tann for a copy of this seminar. It appears in CW 9i as "Concerning Mandala Symbolism" where its Berlin origin is misdated as "1930." Its penultimate paragraph about *Kulturkriese* ("culture zones"), a key concept in the work of Frobenius, is deleted.
24. von der Tann and Erlenmeyer, *C. G. Jung und der Nationalsozialismus*, p. 27. Max Zeller recalled Jung saying "I am afraid that we *had* to let world history pass by." See Janet Dallet, ed., *The Dream—The Vision of the Night* (Los Angeles: Analytical Psychology Club and the C.G. Jung Institute, 1975), p. 128.
25. Letter to *Aufbau* (New York) by Günther Looser, August 26, 1955.
26. *CGJS*, especially pp. 26–28.
27. CW 6, p. 547.
28. CW 11, p. 481.
29. CW 2, pp. 605–614.
30. "An der Schwelle" (Heilbronn: Verlag Eugen Salzer, 1937), p. 86. It is probable that Weizsäcker is the acquaintance Jung mentions as being responsible for alerting him to the liturgical endeavors of the Berneuchner Circle; see *Letters, Volume I*, p. 215. For more on the conference, see Gerhard Wehr, *An Illustrated Biography of C.G. Jung* (Boston and Shaftesbury: Shambhala, 1989), pp. 328–329.
31. CW 11, p. 336.
32. CW 10; and C.G. Jung, *Essays on Contemporary Events* (Kegan Paul: London, 1947).
33. CW10, p. 185. The original reads "dass einer, der offenkundig ergriffen ist, das ganze Volk dermassen ergreift, dass sich alles in Bewegung setzt, ins Rollen gerät und unvermeidlicherweise auch in gefährliches Rutschen."
34. Cocks, *Psychotherapy in the Third Reich*, p. 142.
35. Jung, *Essays on Contemporary Events*, p. 6.
36. Ibid., p. 10.
37. Karl Bracher, *The German Dictatorship* (New York: Praeger, 1976), p. 29. See also Lèon Poliakov, *The Aryan Myth* (New York: Barnes and Noble, 1996), pp. 251–252.

38. Jung, *Essays on Contemporary Events*, p. 11.
39. See *Jahrsbericht* 1936 of the Psychology Club Zurich and "Jung's Work since 1939" by Toni Wolf, both in the Kristine Mann Library. See Margarete Dierks, *Jacob Wihelm Hauer, 1881–1962* (Heidelberg: Verlag Lambert Schneider, 1986), p. 239. In *Glaubenskrise im Dritten Reich* (Stuttgart: Deutsche Verlags Anstalt, 1953), Hans Buchheim noted that under Daur the Bund came to oppose National Socialism and take a position closer to the Confessing Churches (p. 175).
40. Jung, *Essays on Contemporary Events*, p. 11. (*Neue Schweizer Rundschau*, 665/ German CW 10).
41. CW 18, pp. 164.
42. Jung, *Essays on Contemporary Events*, p. 8.
43. See Doris L. Bergin, *Twisted Cross: The German Christian Movement in the Third Reich* (Chapel Hill: University of North Carolina Press, 1996).
44. William Sheridan Allen, ed. and trans., *The Infancy of Nazism: The Memoirs of Ex-Gauleiter Albert Krebs 1923–1933* (New York: New Viewpoints, 1976), p. 287.
45. See Paul Douglass, *God Among the Germans* (Philadelphia: University of Pennsylvania, 1935), pp. 58–72 and *Glaubenskrise*, pp. 164–198.
46. *Letters, Volume I*, p. 212 (March 10, 1936).
47. See Dierks, *Jacob Wilhelm Hauer*, p. 279; and Une Dietrich Adam, *Hochschule und Nationalsozialismus: Die Universitat Tübingen im Dritten Reich* (Tübingen: J.C.B. Mohr, 1977), p. 76.
48. *Letters, Volume I*, p. 233.
49. Dierks, *Jacob Wilhelm Hauer*, pp. 296–297.
50. In his introduction to the Zarathustra Seminar James Jarrett gives an incomplete inventory of Jung's professional activities during this period. For example, he mentions Jung's military obligations but fails to note his far more important presidency of the General Medical Society for Psychotherapy. For a contextualization of the seminar in Nietzsche studies, see Steven Aschenheim, *The Nietzsche Legacy in Germany 1890–1990* (Berkeley: University of California Press, 1992), pp. 258–262.
51. Jolande Jacobi interview, p. 24. Jung Oral History Archive, Countway Medical Library, Boston.
52. William McDougall, *The Group Mind* (New York: Putnam's Sons, 1920), p. 332.
53. *CGJS*, p. 6.
54. *Letters, Volume I*, pp. 195–196.
55. The journal was founded in 1934. In 1935 it had a Carl Häberlin article on Klages and an advertisement for Christian Jenssen's *Deutsche Dichtung der Gegenwart* (Heft 12).
56. See Ulfried Geuter, *The Professionalization of Psychology in Nazi Germany* (Cambridge: Cambridge University Press, 1992), p. 121.
57. *Letters, Volume I*, p. 272.
58. CW 11, p. 335.
59. CW 18, p. 581.
60. E. James Lieberman, *Acts of Will* (New York: Free Press, 1985), p. 379.
61. See also Cocks, *Psychotherapy in the Third Reich*, pp. 40, 330–331.

62. In Hans-Martin Lohman, ed., *Psychoanalyse und Nationalsozialismus* (Frankfurt am Main: Fischer Verlag, 1984), pp. 146–155.
63. For the most comprehensive history of the Frankfurt School, see Martin Jay, *The Dialectical Imagination* (Berkeley: University of California Press, 1996).
64. Gershom Scholem and Adorno W. Theodor, *The Correspondence of Walter Benjamin* (Chicago: University of Chicago Press, 1994), p. 540.
65. Ibid., p. 545.
66. *The Principle of Hope (Vol. I)* (Cambridge: MIT Press, 1996), p. 59.
67. Erich Fromm discussed this issue in "The Theory of Mother Right and Social Psychology," in *The Crisis of Psychoanalysis* (Greenwich, CT: Fawcett Publishers, 1970), pp. 110–135. The article, which originally appeared in the 1934 issue of *Zeitschrift für Sozial Forschung*, discusses the contradictory interpretations of Bachofen's theory of Mother Right. Reactionary intellectuals emphasized its source in the depth of the maternal unconscious while socialists adopted it as the model of a new system of social relations. For more, see Daniel Burston, *The Legacy of Erich Fromm* (Cambridge: Harvard University Press, 1991), pp. 37–45.
68. *The Principle of Hope (Vol. 1)*, p. 63.
69. Herbert Marcuse, *Eros and Civilization* (New York: Vintage, 1962), p. 134ff.
70. See M.E. Warlick, *Max Ernst and Alchemy* (Austin: University of Texas Press, 2001), p. 32.
71. Frederick J. Hoffman, *Freudianism and the Literary Mind* (Baton Rouge: Louisiana State University Press, 1945), p. 45.
72. Letter to James Oppenheim (September 24, 1931). James Oppenheim Collection, New York Public Library.
73. Franz Neumann, *Behemoth* (New York: Harper and Row, 1966), p. 135.
74. Max Weinrich, *Hitler's Professors* (New York: Yiddish Scientific Institute, 1991), p. 247.
75. Christoph Steding, *Das Reich und die Krankheit der Europäischen Kultur* (Hamburg: Hanseatische Verlag, 1938), p. 247.
76. Examples include Nolde and Expressionism; see Elaine S. Hochman, *Architect of Fortune: Mies van der Rohe and the Third Reich* (New York: Weidenfeld & Nicolson, 1989), pp. 165–169.
77. *ZS*, p. 637.
78. For Bion and Beckett's attendance at the Tavistock lecture, see Deirdre Bair, *Samuel Beckett* (New York: Summit Books, 1990), pp. 208–210.
79. CW 18, p. 572.
80. Ibid., p. 575.
81. *CGJS*, pp. 92–93.
82. H.G. Baynes, *Germany Possessed* (London: Jonathan Cape, 1941), p. 13. I bought my copy in a used bookstore in Oxford. It is inscribed with the name "Alan Bullock" so it would seem that the author of *Adolf Hitler: A Study in Tyranny* found some of the initial inspiration for his research in Baynes' book. Bullock received degrees from Wadham College and was later a history fellow at New College.
83. Ibid., p. 12.
84. Forrest G. Robinson, *Love's Story Told* (Cambridge: Harvard University Press, 1992).
85. Claire Douglass, *Translate this Darkness* (New York: Simon & Schuster, 1993); and *Visions, Notes on a Seminar* (Princeton: Princeton University Press, 1997).

86. *Letters, Vol. I*, p. 224.
87. Robinson, *Love's Story Told*, p. 230. Frankfurter had lobbied Harvard's president James Conant to take a firm anti-Nazi stand; see James Hersberg, *James B. Conant* (New York: Knopf, 1993), pp. 87–88, and 96.
88. *Letters, Volume II*, p. xxxv and 1946 letter to Fordham.
89. CW 11, p. 28.
90. *CGJS*, p. 118.
91. CW 18, p. 164.
92. *CGJS*, pp. 132–133; emphasis in the original.
93. *ZS*, p. 813; emphasis in the original. See also Robinson, *Love's Story Told*, p. 124, where Jung used the inevitability of the invasion to justify the morality of an adulterous affair.
94. *Visions Seminar*, pp. 974–975.
95. The most famous case of mass hysteria brought on by these anxieties was Orson Welles' "War of the Worlds" broadcast in 1938. For an analysis of the unconscious symbolism involved in this event, see Jerry Kroth, *Omens and Oracles* (New York: Praeger, 1992). In his book *Psychodrama: A Mental Outlook and Analysis* (Cambridge, MA: Sci-Art Publishers, 1942) Roback quoted a letter from Jung who wrote "Politically I still hold that it would be wiser to let the wolf eat stones until he is lamed by his greed. The loss of a few hitherto independent countries is less horrible than the destruction of European civilization as a whole. Political wisdom, I am afraid, is hard and cruel, but shortsightedness is a hundred times worse" (p. 78).
96. *CGJS*, pp. 181–182; and E.A. Bennet, *Meetings With Jung, Conversations 1946–61* (Zurich: Daimon Verlag, 1991), pp. 126–128. For my previous analyses of this dream, see "Jung, the Jews, and Hitler," in *Spring* 1986, pp. 170–172; and "The Case of Jung's Alleged Anti-Semitism," in *Lingering Shadows*, ed. Aryeh Maidenbaum and Stephen Martin (Boston: Shambhala, 1991), pp. 126–128.
97. *ZS*, pp. 697–698. This relationship also comes out clearly in the following quote: "[Jung] said that he always treated [Princess Marie-Alix Hohenzollern] as a princess, and as a number of her ancestors had had a court astrologer, and placed their trust in that man, she immediately put her trust in him. Jung said that he was playing the historical role of the physician to the King." Jane Cabot Reid, ed., *Jung, My Mother, and I: The Analytic Diaries of Catherine Rush Cabot* (Zurich: Daimon Verlag, 2001), p. 113.
98. *APS*, p. 99.
99. *ZS*, p. 1022.
100. Ibid., pp. 1024–1025.

6 The World War II Years

1. See Christian Leitz, *Sympathy for the Devil: Neutral Europe and Nazi Germany in World War II* (New York: New York University Press, 2001), pp. 10–40; and Adam Labor, *Hitler's Secret Bankers: The Myth of Swiss Neutrality During the Holocaust* (Seacacus, NJ: Birch Lane Press, 1997). The most comprehensive treatment of all aspects of Switzerland's wartime behavior is *Switzerland,*

National Socialism, and the Second World War (Zurich: Pendo Verlag, 2002), which is the final report of the independent commission of experts appointed by the Swiss government.

2. See Aryeh Maidenbaum, "Lingering Shadows: A Personal Perspective," in *Lingering Shadows: Jungians, Freudians, and Anti-Semitism*, ed. Aryeh Maidenbaum and Stephen Martin (Boston: Shambhala, 1991), pp. 297–300.
3. Alan Schom, *Survey of Nazi and Pro-Nazi Groups in Switzerland (1930–45)* (Los Angeles: Simon Weisenthal Center, 1998), pp. 2–3.
4. *Letters, Volume I*, pp. 277–280; and *Handbuch der Schweizer Geschichte (Vol. 2)* (Zurich: Verlag Berichthaus, 1977), p. 1180.
5. ETH Jung Collection, HS 1055:171.
6. *Letters, Volume I*, p. 277.
7. *Schweizer Illustrierte*, August 12, 1942.
8. *MDR*, p. 294.
9. William McGuire, *Bollingen: An Adventure in Collecting the Past* (Princeton: Princeton University Press, 1982), p. 38. See Margaret Chase, ed., *Heinrich Zimmer: Coming into His Own* (Princeton: Princeton University Press, 1994); especially William McGuire, "Zimmer and the Mellons," pp. 31–42.
10. See Hubert Cancik, "Dionysos 1933. W.F. Otto, ein Religionswissenschaftler und Theologe am Ende der Weimarer Republik" in *Die Restauration der Götter: Antike Relgion und Neo-Paganismus*, ed. Richard Faber und Renate Schlesier (Wurzburg: Konigshausen und Neumann, 1986), pgs. 105–23.
11. *Leo Frobenius zum 60th Geburtstag* (Leipzig: Köhler und Amelang, 1933). See also Fritz W. Kramer, "Die Aktualität des Exotischer. Der Fall der 'Kulturmorphologie' von Frobenius und Jensen," in *Die Restauration der Götter.*
12. Remember Jung's swastika anecdote from the 1933 reception at the von Schnitzlers.
13. One of Frobenius' pet theories was that the artifacts he was discovering in West Africa were evidence that the Yoruba culture found there was the basis of the Atlantis myth. See L. Sprague DeCamp, *Lost Continents* (New York: Gnome Press, 1954), pp. 180–184; and James Bramwell, *Lost Atlantis* (Hollywood: Newcastle Publishing, 1974), pp. 119–122. Remember Jung's affection for H. Rider Haggard's *She* and Pierre Benoit's *L'Atlantide.*
14. Grandduke Heinrich at Darmstadt (Library of Congress Jung Archive file, 2:7 (107, handwritten).
15. See Heather Pringle, *The Master Plan: Hitler's Scholars and the Holocaust* (New York: Hyperion, 2006), pp. 102–120.
16. Pantheon Ak. Verlag in Amsterdam/Leipzig.
17. Ernest Harms file, Kristine Mann Library, New York.
18. Centre Documentation Juive Contemporaine (Paris), Document B 6602. My thanks to Martin Galland for generously making a copy of this document available to me.
19. *Simon Wiesenthal Report*, p. 4.
20. For Jung's response, see *Letters, Volume I*, pp. 340–341.
21. Ibid., pp. 328–329.
22. Schom, *Simon Wiesenthal Report*, p. 6.

23. See also "Psych. Disputation," in *Basler National Zeitung*, February 17, 1944, for leftist criticism of Jung's appointment.
24. CW 15, p. 15.
25. *Letters, Volume I*, p. 328.
26. Ibid., p. 331.
27. CW 9i, p. 264.
28. *CGJS*, p. 194.
29. *Letters. Volume I*, p. 154.
30. Ibid., p. 233.
31. *CGJS*, p. 88.
32. Ibid., p. 93. In a wartime letter (July 26, 1942) Carl Burckhardt wrote to Max Rychner about a rendezvous he had with Jung in Zurich. After discussing his *Psychology and Religion* and Nostradamus Jung then brought up Roosevelt "the limping messenger of the Apocalypse who like all Americans has a huge mother-complex and who is an idealist in order to please his mother and to please the Woman in the mother he must be a pimp and a dick." *The Carl Burckhardt-Max Rychner Letters 1926–1965*, ed. Claudia Mertz-Rychner (Frankfurt am Main: Fischer Verlag, 1970), p. 78. I want to thank Wedigo de Vivanco and Florian Galler for their help in translating this letter.
33. CW 9i, p. 374.
34. CW 13, p. 250.
35. In *Hitler, Diagnosis of a Destructive Prophet* (New York: Oxford University Press, 1999) the author Dr. Fritz Redlich refers to his general indebtedness to Jung among others but does not avail himself of the psychiatric judgments found in Jung's pre- and postwar interviews. Jung's name does not even appear in the index of Ron Rosenbaum's *Explaining Hitler* (New York: Random House, 1998).
36. CW 18, p. 604.
37. *CGJS*, p. 140.
38. *CGJS*, p. 127.
39. John Toland, *Adolf Hitler* (New York: Doubleday, 1976), p. 525.
40. Ernst Hanfstaengl, *Hitler: The Missing Years* (New York: Arcade Publishing, 1994), pp. 292–296.
41. Walter C. Langer, *The Mind of Adolf Hitler: The Secret Wartime Report* (New York: New American Library, 1972).
42. Forrest Robinson, *Love's Story Told: A Life of Henry Murray* (Cambridge: Harvard University Press, 1992), pp. 276–278 and f.n. 428.
43. Robert Waite, *The Psychopathic God: Adolf Hitler* (New York: Basic Books, 1977). Waite cites Erik Erikson as a source of advice for his research. For Erikson's involvement with the psychoanalysis of Hitler, see Lawrence J. Friedman, *Identity's Architect: A Biography of Erik Erikson* (New York: Scribner's, 1999), pp. 163–176.
44. Peter Grose, *Gentleman Spy: The Life of Allen Dulles* (Boston: A Richard Todd Book, 1994), p. 163.
45. Mary Bancroft, *Autobiography of a Spy* (New York: William Morrow & Co, 1983), p. 171. See also Joan Dulles Buresch-Talley, "The C.G. Jung and Allen

Dulles Correspondence," in ed. Aryeh Maidenbaum, *Jung and the Shadow of Anti-Semitism* (Berwick, ME: Nicolas Hayes, Inc., 2002), pp. 39–54.
46. Ibid., p. 171.
47. Ibid., p. 191.
48. Hans Gisevius, *To the Bitter End* (Boston: Houghton Mifflin Co., 1947). See also Joseph E. Persico, *Piercing the Reich* (New York: Viking, 1979), chapter IV, "Back Door to the Reich."
49. Quoted in Christopher Simpson, *Splendid Blond Beast* (Monroe, ME: Common Courage Press, 1995), p. 122. See also Agostino von Hassell and Sigrid MacRae, *Alliance of Enemies* (New York: Thomas Dunne Books, 2006), for a comprehensive account of these contacts.
50. Simpson, *Splendid Blond Beast*, p. 156.
51. Ibid., p. 84.
52. Joseph Borkin, *The Crime and Punishment of I. G. Farben* (New York: Free Press, 1978), p. 154.
53. Ferdinand Sauerbruch, *A Surgeon's Life* (London: Andre Deutsch Publishing, 1953), p. 217.
54. See Max Weinrich, *Hitler's Professors* (New York: Yiddish Scientific Institute, 1991), p. 14 and Victor Farias, *Heidegger and Nazism* (Philadelphia: Temple University Press, 1989), p. 156. For more on Hirsch, see Robert Ericksen, *Theology Under Hitler* (New Haven: Yale University Press, 1985).
55. From "Jung and his Circle," by Mary Bancroft, *Psychological Perspectives* (Volume 6, No. 2, 6:2), pp. 123–124.
56. The rumors went back five years. Jung posed the same question to Henry Murray in 1938; see *Letters, Volume II*, pp. xxxiv–xxxv.
57. Sauerbruch, *A Surgeon's Life*, pp. 235, 246–247.
58. See Geoffrey Cocks, *Psychotherapy in the Third Reich* (New Brunswick: Transaction, 1997), pp. 243–244; and Wilhelm Bitter file in Jung Oral History Archive, Countway Medical Library, Boston. In his footnote Cocks corrects Hannah's "1942" to "1943."
59. Grose, *Gentleman Spy*, p. 165.
60. Bancroft, *Autobiography of a Spy*, p. 173.
61. *Letters, Volume I*, p. 357.
62. CIA Document ID: 37376.
63. Paul Mellon, *Reflections in a Silver Spoon: A Memoir* (New York: William Morrow, 1992), p. 219.

7 The Cold War Years

1. *CGJS*, pp. 149–155.
2. Ibid., p. 149.
3. Ibid., p. 154.
4. Ibid., pp. 154–155; see also CW 18, pp. 591–603.

5. See 1923 letter to Oscar A.H. Schmitz.
6. See Jung's letter of December 31, 1949 (*Letters, Volume I*, pp. 539–541). Their complete correspondence is available in Ann C. Lammers and A. Cunningham, eds., *The Jung-White Letters* (London and New York: Routledge [Philemon Series], 2007).
7. CW10, p. 195; emphasis in the original. The most well-known story of Jung's private admission of error was his telling Rabbi Leo Baeck after the war that he had "slipped up." See William McGuire, *Bollingen: An Adventure in Collecting the Past* (Princeton: Princeton University Press, 1982), pp. 152–153.
8. Aryeh Maidenbaum and Stephen Martin, eds., *Lingering Shadows* (Boston: Shambhala, 1991), p. 129.
9. CW 10, p. 207.
10. *Mythology and Humanism: The Correspondence of Thomas Mann and Karl Kerenyi* translated by Alexander Gelley (Ithaca: Cornell University Press, 1975), p. 146; emphasis in the original.
11. CW 10, p. 214.
12. Ibid., p. 210.
13. Ibid., p. 199.
14. Ibid., p. 205.
15. Ibid., p. 209. For additional material on Jung's diagnosis. see "Answers to 'Mishmar' on Adolph Hitler" (CW 18, pp. 604–605).
16. CW 10, p. 539.
17. Ibid., p. 236.
18. Ibid., p. 166.
19. Ibid., p. 237.
20. CW 17, p. 185.
21. Jung letter to Fordham (April 18, 1946). I want to thank Sonu Shamdasani for making this letter available to me.
22. ETH Jung Archive, 1055: 995, p. 49 (dated January 1946) (ZBB: 26–27).
23. Gerhard Adler, *Dynamics of the Self* (London: Coventure, 1979), p. 99. For a more recent, non-Jungian interpretation, see "'A Little Sun in His Own Heart': The Melancholic Vision in *Answer to Job*," in *Men, Religion, and Melancholia*, ed., Donald Capp (New Haven: Yale University Press: 1997), pp. 127–151.
24. See Ralph Metzner, *The Well of Remembrance* (Boston: Shambhala, 1994), pp. 20–21.
25. William Schoenl, *C.G. Jung: His Friendship with Mary Mellon and J.B. Priestly* (Wilamette, IL: Chiron, 1998), pp. 33–36.
26. McGuire, *Bollingen*, p. 72.
27. This document was sent to me by the CIA under the Freedom of Information Act (ID #37375).
28. Barbara Hannah, *Jung: His Life and Work* (New York: Putnam and Sons, 1976), p. 289, f.n. "b."
29. *Letters, Volume I*, p. 406.
30. Regine Lockot, *Die Reinigung der Psychanalyse* (Tübingen: edition discord, 1994), p. 119 (translation by Charles Boyd).
31. *Letters, Volume I*, p. 445.
32. Ibid., pp. 424–425.

33. Geoffrey Cocks, *Psychotherapy in the Third Reich* (New Brunswick: Transaction, 1997), p. 184.
34. Alex von Muralt, *Schweizer Annalen* (No. 12, 1947), pp. 672–702, reprinted in *Hamburger Akademische Rundschau* (Heft, No. 7, 1948/49, pp. 546–557).
35. See "Für und Gegen C.G. Jung," in *Aufbau* (New York City) April 27, 1956.
36. For Röpke and Schönbrunner, see Alexander and Selesnick, *History of Psychiatry* (New York: Harper and Row, 1964), Appendix B.
37. *APC of NY Bulletin* (June 1946), p. 4.
38. Harms letter about NY Jungians in Ernest Harms Collection, Kristine Mann Library.
39. McGuire, *Bollingen*, p. 213.
40. August–September, p. 29.
41. February–March, p. 23.
42. See *Letters, Volume I*, pp. 479–480; *Letters, Volume II*, pp. xxvii–xxxix.
43. *The New Republic* (December 4, 1944), p. 744.
44. Paul Mellon, *Reflections in a Silver Spoon: A Memoir* (New York: William Morrow, 1992), pp. 171–177. "It is however difficult to mention the anti-christianism of the Jews after the horrible things that have happened in Germany. But Jews are not so damned innocent after all—the role played by the intellectual Jews in prewar Germany would be an interesting object of investigation" (letter to Mary Mellon September 24, 1945, in William Schoenl, *C.G. Jung: His Friendship with Mary Mellon and J.B. Priestley* (Wilmette, IL: Chiron Press, 1998), p. 39.
45. See McGuire, *Bollingen*, pp. 123f.
46. Bollingen Collection, Library of Congress (B–I: 99).
47. Norman F. Cantor, *Inventing the Middle Ages* (New York: Quill William Morrow, 1991), p. 212.
48. *The New Yorker* (December 14, 1946), p. 36.
49. Meridian Books (Cleveland and New York: World Publishing Co., 1963), p. 148. Based on this critique, which "makes further discussion of Jung's work unnecessary," Herbert Marcuse dismissed Jung's psychology for its "obscurantistic and reactionary trends" (*Eros and Civilization* [New York: Vintage, 1962], p. 134 and f.n.).
50. See Marvin Goldwert, "Toynbee and Jung: the Historian and the Analytical Psychologist—A Brief Encounter," *Journal of Analytical Psychology* (Vol. 28, 1983), pp. 363–366.
51. For Jung, see *CGJS*, p. 93; for Toynbee's views, see William McNeil, *Toynbee: A Life* (Oxford: Oxford University Press, 1989), p. 218.
52. McNeil, *Toynbee: A Life*, p. 218.
53. Ernest Jones, *The Life and Works of Sigmund Freund, Vol. 2* (New York: Basic Books, 1955), pp. 102–103.
54. Paul Roazen, *Freud and His Followers* (New York: Knopf, 1976), p. 262.
55. CW 10, p. 114; emphasis in the original.
56. Wilhelm Bitter, *Die Krankheit Europas* (Freiburg im Bresigau: Niels Kampfmann, 1932), p. 33.
57. Quoted in *Gentleman Spy*, p. 269.
58. See Tom Bower, *The Paperclip Conspiracy* (New York: Little Brown, 1987) and Wayne Biddle, *The Dark Side of the Moon: Werner Von Braun, the Third Reich, and the Space Race* (New York: Norton, 2009).

59. Pied Piper (fn 58) *Letters Vol. II*, pp. 330–32.
60. Peter Grose, *Gentleman Spy: The Life of Allen Dulles* (Boston: Richard Todd, 1994), p. 321. See also Frances Stonor Saunders, *The Cultural Cold War: The CIA and the World of Arts and Letters* (New York: Free Press, 1999); and Volker Berghahn, *America and the Intellectual Cold Wars in Europe* (Princeton: Princeton University Press, 2001), pp. 216–218.
61. Koestler, *Letters, Volume II,* pp. 600–603.
62. Dulles-Mary Bancoft in Grose, *Gentleman Spy*, p. 430.
63. *Foreign Affairs* reference in "Immer der Flucht vor Sicht die Deutschen," *Die Welt* (Hamburg), April 5, 1958.
64. CW 3, p. 250. "Recent Thoughts on Schizophrenia," originally in the *Bulletin of the APC of New York* (Volume 19 No. 4, April 1957).
65. CW 18, p. 636.
66. *Letters, Volume I*, p. 505.
67. *Letters, Volume II*, p. 85.
68. CW 18, p. 609.
69. Peter Homans, *Jung in Context* (Chicago: The University of Chicago Press, 1995), pp. 179–182.
70. Ibid., p. 181.
71. Jung Book Catalog (1967): *Mensch und Geschichte* (1929) and *Die Stellung des Mensch im Kosmos* (1928). The latter was Scheler's address at the 1927 conference of the School of Wisdom. It is available in English as *Man's Place in Nature* (New York: Noonday Press, 1971).
72. Jung in *Revista de Occidente*: "Psychological Types" (1925); *The Unconscious in Normal and Pathological Mind* (1927); "Archaic Man" (1931); "The Psychological Problems of Modern Man (1932); "Ulysses" (1933); "Picasso" (1934); "The Archetypes of the Collective Unconscious" (1936) (CW 19, pp. 143–144). In "Phrase and Sincerity' (*Europäische Revue*, 1/34) Ortega y Gasset mentions Jung's use of the "Paleo-ontology of the Soul." He admits that he could not find the usage in Jung's works but said that it was a formulation that Jung used in conversation with Count Keyserling (p. 26). It is possible that it originated with Jelliffe who spoke about "paleopsychology [see John Burnham and William McGuire, *Jelliffe: American Psychoanalyst and Physician and His Correspondence with Sigmund Freud and C.G. Jung.* (Chicago: University of Chicago Press, 1983), pp. 243–245].
73. Alan Schom, *Simon Wiesenthal Report*, p. 6. Jung continued to contribute the occasional article to the affiliated *Zürcher Student*; see "The Effect of Technology on the Human Psyche" (1949; CW 18, pp. 614–615) and "On Psychodiagnostics" (1958; CW 18, p. 637).
74. Another link, albeit indirect, that Jung maintained to the Swiss Right was his employment of Argus, an international press-cutting service. It was owned by Rolf Henne, the young relative who had been active in prewar fascist activities. See "The National Front," in *Who Were the Fascists: Social Roots of European Fascism*, ed. Stein Ugelvik Larsen, Bernt Hagtvet, and Jan Peter Myklebust (Bergen, Oslo, Tromsph [o/]: Universitets Forlaget, 1980), p. 478, f.n. 18.
75. CW 10, p. 247.
76. Ibid., p. 264.
77. Ibid., p. 267.

78. Quoted in A. Scott Berg, *Lindbergh* (New York: Putnam & Sons, 1998), p. 511.
79. *CGJS*, p. 395.
80. See Sonu Shamdasani, *Jung Stripped Bare By His Biographers, Even* (London: Karnac Books, 2005), pp. 22–38.
81. Serrano's book is *Adolf Hitler, el Ultimo Avatara* (Bogota: Editorial Solar, 2000). In a letter to Serrano (September 14, 1960) Jung returned to his Wotan hypothesis when he wrote "As we have largely lost our Gods and the actual condition of our religion does not offer a efficacious answer to the world situation in general and to the 'religion' of communism in particular, we are very much in the same predicament as the pre-National-Socialistic Germany of the Twenties, i.e. we are apt to undergo the risk of a further, but this time worldwide, Wotanistic experiment. This means mental epidemy and war." In Miguel Serrano, *C.G. Jung and Hermann Hesse: A Record of Two Friendships* (Schocken Books: New York, 1966), p. 85. For more on Serrano, see Joscelyn Goodwin, *Arktos: The Polar Myth in Science, Symbolism, and Nazi Survival* (Grand Rapids, MI: Phanes Press, 1993), pp. 70–73.
82. *Letters, Volume II*, p. 298.
83. Angela Graf-Nolde, "C. G. Jung's Position at the Eidgenössische Technische Hochschule Zürich (ETH Zurich): The Swiss Federal Institute of Technology, Zurich," in *Jung History* (Volume 2, No. 2, Fall 2007), pp. 12–15.

Bibliography

Abraham, Hilda, and Ernst L. Freud, eds., *A Psycho-Analytical Dialogue: The Letters of Sigmund Freud Abraham 1907–1926.* New York: Basic Books, 1965.

Adam, Une Dietrich. *Hochschule und Nationalsozialismus: Die Universität Tübingen im Dritten Reich.* Tübingen: J.C.B. Mohr, 1977.

Adler, Gerhard. *Dynamics of the Self.* London: Coventure, 1979.

Allen, William Sheridan, ed. and trans. *The Infancy of Nazism: The Memoirs of Ex-Gauleiter Albert Krebs 1923–1933.* New York: New Viewpoints, 1976.

Aschenheim, Steven. *The Nietzsche Legacy in Germany, 1890–1990.* Berkeley: University of California Press, 1992.

Bair, Deirdre. *Samuel Beckett.* New York: Summit, 1990.

Bancroft, Mary. *Autobiography of a Spy.* New York: William Morrow, 1983.

———. "Jung and his Circle." *Psychological Perspectives* 6:2 (Fall 1985): 114–127.

Baudouin, Charles. *L'Oeuvre de Jung.* Paris: Payot, 1963.

Baumgart, Ursula. "Betrachtung eines Bildes von C.G. Jung." *Analytische Psychologie* 18 (1987): 303–312.

Baynes, H.G. *Germany Possessed.* London: Jonathan Cape, 1941.

Begg, Ian. "Jung's Lost Lieutenant." *Harvest* 22 (1976): 78–90.

Bennet, E.A. *Meetings With Jung, Conversations 1946–61.* Zurich: Daimon Verlag, 1991.

Bentley, Toni. *Sisters of Salome.* New Haven: Yale University Press, 2003.

Berg, A. Scott. *Lindbergh.* New York: Putnam, 1998.

Berghahn, Volker. *America and the Intellectual Cold Wars in Europe.* Princeton: Princeton University Press, 2001.

Bergin, Doris L. *Twisted Cross: The German Christian Movement in the Third Reich.* Chapel Hill: University of North Carolina Press, 1996.

Bernfeld, Suzanne Cassirer. "Freud and Archeology." *American Imago* 8:2 (1951): 107–128.

Bilski, Emily, ed. *Golem: Danger, Deliverance, and Art.* New York: The Jewish Museum, 1988.

Binswanger, Ludwig. *Sigmund Freud: Reminiscences of a Friendship.* New York and London: Grune & Stratton, 1957.

Bishop, Paul. *The Dionysian Self: C.G. Jung's Reception of Friedrich Nietzsche.* Berlin: de Gruyter, 1995.

Bitter, Wilhelm. *Die Krankheit Europas.* Freiburg im Bresigau: Niels Kampfmann, 1932.

Bloch, Ernst. *The Principle of Hope.* Cambridge: MIT Press, 1996.
Blüher, Hans. *Werke und Tage.* München: Paul List Verlag, 1953.
Boe, John. "Jack London, The Wolf, and Jung." *Psychological Perspectives* 11:2 (1980): 133–136.
Bon, Gustave Le. *Psychology of the Great War.* New Brunswick: Transaction, 1999.
Borkin, Joseph. *The Crime and Punishment of I. G. Farben.* New York: Free Press, 1978.
Brabant, Eva, Ernst Falzeder, Patrizia Giampieri-Deutsch, eds. and trans. *The Correspondence of Sigmund Freud and Sandor Ferenczi, Volume One 1908–1914.* Cambridge: Belknap Press, 1993.
Bracher, Karl. *The German Dictatorship.* New York: Praeger, 1976.
Bramwell, James. *Lost Atlantis.* Hollywood: Newcastle Publishing, 1974.
Brenner, Michael. *The Renaissance of Jewish Culture in Weimar Germany.* New Haven: Yale University Press, 1996.
Brockway, Robert. *Young Carl Jung.* Wilmette, IL: Chiron, 1996.
Brode, Hanspeter. *Benn Chronik.* München: Carl Hanser Verlag, 1978.
Brome, Vincent. *Ernest Jones: Freud's Alter Ego.* New York: Norton, 1983.
Buchheim, Hans. *Glaubenkrise im Dritten Reich.* Stuttgart: Deutsche Verlags Anstalt, 1993.
Buchholz, Kai, Rita Latocha, Hilke Peckmann, Klaus Wolbert, ed. *Die Lebensreform: Entwürfe zur Neugestaltung von Leben und Kunst um 1900.* Darmstadt: Institut Mathildenhöhe [Verlag Häusser], 2003.
Burckhardt, Jacob. *Reflections on History.* Indianapolis: Liberty Classics, 1979.
———. *The Greeks and Greek Civilization.* New York: St. Martin's Press, 1998.
Buresch-Talley, Joan Dulles. "The C.G. Jung and Allen Dulles Correspondence." In *Jung and the Shadow of Anti-Semitism*, edited by Aryed Maidenbaum, 39–54. Berwick, ME: Nicolas-Hays, 2002.
Burnham, John, and William McGuire. *Jelliffe: American Psychoanalyst and Physician and His Correspondence with Sigmund Freud and C.G. Jung.* Chicago: University of Chicago Press, 1983.
Burston, Daniel. *The Legacy of Erich Fromm.* Cambridge: Harvard University Press, 1991.
Campbell, Joseph. "Introduction." In *Myth, Religion, and Mother Right*, edited by J.J. Bachofen, xxv–lvii. Princeton: Princeton University Press, 1973.
Cancik, Hubert. "Dionysos 1933. W.F. Otto, ein Religionswissenschaftler und Theologe am Ende der Weimarer Republik." In *Die Restauration der Gotter: Antike Relgion und Neo-Paganismus*, edited by Richard Faber and Renate Schlesier, 105–123. Wurzburg: Konigshausen und Neuman, 1986.
Cantor, Norman F. *Inventing the Middle Ages.* New York: Quill William Morrow, 1991.
Capp, Donald. *Men, Religion, and Melancholia.* New Haven: Yale University Press, 1994.
Carotenuto, Aldo. *A Secret Symmetry.* New York: Pantheon, 1982.
Carus, Carl Gustav. *Nine Letters on Landscape Painting.* Los Angeles: Getty Research Institute, 2002.
Cavendish, Richard. *Encyclopedia of the Unexplained.* New York: McGraw-Hill, 1974.
Chamberlain, Houston Stuart. *Foundations of the Nineteenth Century.* London: Bodley Head, 1914.

Charet, F.X. *Spiritualism and the Foundation of C.G. Jung's Psychology*. Albany: State University of New York Press, 1993.

Chase, Margaret, ed. *Heinrich Zimmer: Coming into His Own*. Princeton: Princeton Universsity Press, 1994.

Cocks, Geoffrey. *Psychotherapy in the Third Reich*. New Brunswick: Transaction, 1997.

Dallet, Janet, ed. *The Dream—The Vision of the Night*. Los Angeles: Analytical Psychology Club and the C.G. Jung Institute, 1975.

Davidson, Robert F. "Rudolf Otto's Interpretation of Religion." *The Review of Religion* (November 1940): 46–66.

Deak, Frantisek. *Symbolist Theater: The Formation of an Avant-Garde*. Baltimore: Johns Hopkins University Press, 1993.

DeCamp, L. Sprague. *Lost Continents*. New York: Gnome Press, 1954.

Die Kulturelle Bedeutung Der Komplexen Psychologie. Berlin: Julius Springer Verlag, 1935.

Diercks, Margarete. *Jacob Wilhelm Hauer 1881–1962*. Heidelberg: Verlag Lambert Schneider, 1986.

Dijkstra, Bram. *Idols of Perversity*. Oxford: Oxford University Press, 1986.

Donn, Linda. *Freud and Jung: Years of Friendship, Years of Loss*. New York: Scribner's, 1988.

Douglass, Claire. *Translate this Darkness: The Life of Christiana Morgan, the Veiled Woman in Jung's Circle*. New York: Simon & Schuster, 1993.

Douglass, Paul. *God Among the Germans*. Philadelphia: University of Pennsylvania Press, 1935.

Dru, Alexander, ed. and tr. *The Letters of Jacob Burckhardt*. London: Routledge, & Kegan Paul, 1955.

Edwards, Holly, et al. *Noble Dreams, Wicked Pleasures: Orientalism in America 1870–1930*. Princeton: Princeton University Press/Clark Art Institute, 2000.

Efron, John. *Defenders of the Race: Jewish Doctors & Race Science in Fin-de-Siècle Europe*. New Haven: Yale University Press, 1994.

Ellenberger, Henri. *The Discovery of the Unconscious*. New York: Basic Books, 1970.

———. "C.G. Jung and the Story of Helene Preiswerk: A Critical Study with New Documents." In *Beyond the Unconscious*, edited by Mark Micale, 291–305. Princeton: Princeton University Press, 1993.

Ellwood, Robert. *The Politics of Myth*. Albany: State University of New York Press, 1999.

Ermath, Michael ed. *Kurt Wolff: A Portrait in Essays and Letters*. Chicago: University of Chicago Press, 1991.

Etherington, Norman, ed. *The Annotated She*. Bloomington: Indiana University Press, 1991.

Farias, Victor. *Heidegger and Nazism*. Philadephia: Temple University Press, 1989.

Faye, Emmanuel. Michael B. Smith, trans. *Heidegger, The Introduction of Nazism into Philosophy*. New Haven: Yale University Press, 2009.

Field, Geoffrey. *The Evangelist of Race*. New York: Columbia University Press, 1981.

Fischer, Klaus. *Nazi Germany: A New History*. New York: Continuum, 1995.

Foote, Edward. "Who was Mary Foote?" *Spring 1974* 256–268.

Fordham, Michael. "Memories and Thoughts about C.G. Jung." In *Freud, Jung, Klein: The Fenceless Field*, edited by Sonu Shamdasani, 107–113. London: Routledge, 1995.

Forth, Christopher E. *The Dreyfus Affair and the Crisis of French Manhood*. Baltimore: The Johns Hopkins University Press, 2004.

Franz, Marie Louise von. *Psychotherapy*. Boston: Shambhala, 1993.

Freeman, Ralph. *Herman Hesse: Pilgrim of Crisis*. New York: Pantheon, 1978.

Freud, Sigmund. *On the History of the Psycho-Analytic Movement*. New York: Norton, 1966.

Fromm, Erich. *The Crisis of Psychoanalysis*. Greenwich, CT: Fawcett, 1970.

Fuchs, Georg. *Sturm und Drang in München*. München: Verlag Callwey, 1936.

Gasman, Daniel. *The Scientific Origins of National Socialism*. London: Mac Donald; New York: American Elsevier, 1971.

Gelley, Alexander, tr. *Mythology and Humanism: The Correspondence of Thomas Mann and Karl Kerenyi*. Ithaca: Cornell University Press, 1975.

Gerlick, Rhonda. *Electric Salome: Loie Fuller's Performance of Modernism*. Princeton: Princeton University Press, 2007.

German Fiction Writers: 1885–1913 "[Dictionary of Literary Biography, Volume 66]. Detroit: B. C. Layman, 1988.

Geuter, Ulfried. *The Professionalization of Psychology in Nazi Germany*. Cambridge: Cambridge University Press, 1992.

Gildea, Margaret. "Jung As Seen By An Editor, A Student, And A Disciple,." In *Carl Jung, Emma Jung, and Toni Wolff*, 23–27. San Francisco: Analytical Psychology Club of San Francisco, 1982.

Gilman, Sander. *Freud, Race, and Gender*. Princeton: Princeton University Press, 1993.

Gisevius, Hans. *To the Bitter End*. Boston: Houghton Mifflin, 1947.

Gold, Arthur, and Robert Fitzdale. *The Divine Sarah*. New York: Knopf, 1991.

Goldwert, Marvin. "Toynbee and Jung: the Historian and the Analytical Psychologist—A Brief Encounter." *Journal of Analytical Psychology* 28 (1983): 363–366.

Goodwin, Joscelyn. *Arktos: The Polar Myth in Science, Symbolism, and Nazi Survival*. Grand Rapids, MI: Phanes Press, 1993.

Gosset, Thomas. *Race*. New York: Oxford University Press, 1997.

Gossman, Lionel. "Basle, Bachofen, and the Critique of Modernity in the Second Half of the Nineteenth Century." *Journal of the Warburg and Courtauld Institutes* 47 (1984): 141.

———. *Basel in the Age of Burckhardt*. Chicago: University of Chicago Press, 2000.

Graf-Nolde, Angela. "C. G. Jung's Position at the Eidgenössische Technische Hochschule Zürich (ETH Zurich): The Swiss Federal Institute of Technology, Zurich." *Jung History* 2:2 (2007): 12–15.

Gray, Richard. *About Face: German Physiognomic Thought from Lavater to Auschwitz*. Detroit: Wayne State University Press, 2004.

Green, Martin. *The von Richthofen Sisters*. New York: Basic Books, 1974.

———. *Mountain of Truth: The Counter Culture Begins, Ascona 1900–1920*. Hanover, NH: University Press of New England, 1986.

———. *Otto Gross, Freudian Psychoanalyst 1877–1920.* Lewiston, NY: The Edwin Mellen Press, 1999.

Gregory, Frederick. *Nature Lost: Natural Science and the German Theological Traditions of the Nineteenth Century.* Cambridge: Harvard University Press, 1992.

Grose, Peter. *Gentleman Spy: The Life of Allen Dulles.* Boston: Richard Todd, 1994.

Grossman, S. "C.G. Jung and National Socialism." *Journal of European Studies* ix (1979): 231–259.

Hakl, Hans Thomas. *Der verborgene Geist von Eranos.* Bretten: Verlag Neue Wissenschaft, 2001.

Hale, Nathan. *The Rise and Crisis of Psychoanalysis in the U.S.: Freud and the Americans 1917–85.* New York: Oxford University Press, 1992.

Handbuch der Schweizer Geschichte. Zurich: Verlag Berichthaus, 1977.

Hanfstaengl, Ernst. *Hitler: The Missing Years.* New York: Arcade Publishing, 1994.

Hannah, Barbara. *Jung: His Life and Work.* New York: Putnam and Sons, 1976.

Harrington, Anne. *Reenchanted Science: Holism in German Culture from Wilhelm II to Hitler.* Princeton: Princeton University Press, 1996.

Harrowitz, Nancy, ed. *Tainted Greatness: Antisemitism and Cultural Heroes.* Philadelphia: Temple University Press, 1994.

Hayman, Ronald. *A Life of Jung.* London: Bloomsbury Press, 1999.

Heisig, James. "The VII Sermones: Play and Theory." *Spring 1972* 206–18.

Herf, Jeffrey. *Reactionary Modernism: Technology, Culture, and Politics Weimar and the Third Reich.* New York: Cambridge University Press, 1984.

Hersberg, James. *James B. Conant.* New York: Knopf, 1993.

Heuer, Gottfried. "Jung's Twin Brother. Otto Gross and Carl Gustav Jung." *Journal of Analytical Psychology* 46:3 (2001): 655–688.

Hillman, James. "Some Early Background to Jung's Ideas: Notes on C.G. Jung's Medium by Stephanie Zumstein-Preiswerk." *Spring 1976* 123–136.

Hinde, John. *Jacob Burckhardt and the Crisis of Modernity.* Montreal and Kingston: McGill-Queen's University Press, 2000.

Hirsch, Susan. "Hodler as Genevois, Hodler as Swiss." In *Ferdinand Hodler, Views and Visions*, edited by Juerg Albrecht and Peter Fischer, 66–107. Zurich: Swiss Institute for Art Research, 1994.

Hochman, Elaine S. *Architect of Fortune: Mies van der Rohe and the Third Reich.* New York: Weidenfeld & Nicolson, 1989.

Hoffman, Frederick J. *Freudianism and the Literary Mind.* Baton Rouge: Louisiana State University Press, 1945.

Holmes, Stephen. *The Anatomy of Antiliberalism.* Cambridge: Harvard University Press, 1993.

Homans, Peter. *Jung in Context.* Chicago: The University of Chicago Press, 1995.

Horne, John, and Alan Kramer. *German Atrocities 1914: A History of Denial.* New Haven: Yale University Press, 2001.

Howard, Thomas Albert. *Religion and the Rise of Historicism.* Cambridge: Cambridge University Press, 2000.

In the Footsteps of Goethe. Düsseldorf: C.G Bornier, 1999.

Jaffe, Aniela, ed. *Word and Image.* Princeton: Princeton University Press, 1979.

———. "Details about C.G. Jung's Family." *Spring 1984* 35–43.

Jay, Martin. *The Dialectical Imagination.* Berkeley: University of California Press, 1996.

Jenkins, Paul. *"CMS' Early Experiment in Inter-European Cooperation".* Church Missionary Society and the World Church 1799–1999. Basel: Basel Mission, 1999.

Jones, Ernest. *Four Centenary Addresses.* New York: Basic Books, 1956.

———. *The Life and Work of Sigmund Freud.* New York: Basic Books, 1955.

Jones, Russell. "Psychology, History, and the Press, The Case of William McDougall and the New York Times." *American Psychologist* 42:10 (1987): 931–940.

Jung, C.G. *Contributions to Analytical Psychology.* New York: Harcourt, Brace, 1928.

———. *Two Essays on Analytical Psychology.* New York: Dodd, Mead, 1928.

——— . *Modern Man in Search of a Soul.* New York: Harcourt Brace, 1933.

———. "Farewell Address." *Kristine Mann Library file.* New York, October 1936.

———. *Psychology of the Unconscious.* New York: Dodd, Mead, 1946.

———. *Essays on Contemporary Events.* London: Kegan Paul, 1947.

———. *The Undiscovered Self* (Atlantic Monthly Press Book). Boston: Little, Brown, 1958. (Now in CW 10.)

———. *Flying Saucers: A Modern Myth of Things Seen in the Sky.* New York: Harcourt Brace, 1959. (Now in CW 10.)

———. *Man and His Symbols.* Garden City: Doubleday, 1964.

———. *Letters.* 2 Volumes. Princeton: Princeton University Press, 1973.

———. *C.G. Jung Speaking: Interviews and Encounters.* Ed. William McGuire and R.F.C. Hull. Princeton: Princeton University Press, 1977.

———. *The Zofingia Lectures.* Princeton: Princeton University Press, 1983.

———. *Dream Analysis Seminar.* Princeton: Princeton University Press, 1984.

———. *Erinnerungen, Träume, Gedanken.* Olten/Freiburg im Breisgau: Walter Verlag, 1987.

———. *Zarathustra Seminar.* Princeton: Princeton University Press, 1988.

———. *Memories, Dreams, Reflections.* New York: Vintage Books, 1989.

———. *Analytical Psychology: Notes on the Seminar Given in 1925.* Princeton: Princeton University Press, 1989.

———. *The Psychology of Kundalini Yoga*, ed. Sonu Shamdasani. Princeton: Princeton University Press, 1996.

———. *Visions Seminar.* Princeton: Princeton University Press, 1997.

———. *The Collected Works,* ed. Herbert Read, et al., trans. R.F.C. Hull. 20 Volumes. Princeton: Princeton University Press, 1953–1979.

———. *The Red Book*, ed. Sonu Shamdasani. New York: W.W. Norton, 2009.

Jung, Edgar. *Die Sinndeutung der deutschen Revolution.* Oldenburg: Stalling, 1933.

Kaes, Anton, Jay Martin, and Edward Dimenberg,. *The Weimar Republic Sourcebook.* Berkeley: University of California Press, 1995.

Kahan, Alan. *Aristocratic Liberalism.* New Brunswick, NJ: Transaction, 2001.

Kandinsky, Wassily. *Concerning the Spiritual in Art.* New York: Dover, 1977.

Kerr, John. *A Most Dangerous Method.* New York: Knopf, 1993.

Kershaw, Alex. *Jack London: A Life.* New York: St. Martin's Press, 1997.

Kessler, Harry. *Berlin in Lights: The Diaries of Count Harry Kessler, 1918–37.* New York: Grove Press, 1999.

Keyserling, Hermann. *Europe.* New York: Harcourt, Brace, 1928.

———. *The Recovery of Truth.* New York: Harper's, 1929.

Kindler, Nina. "G.R. Heyer in Deutschland." In *Die Psychologie des 20. Jahrhunderts, Band III: Freud und die Folgen*, edited by Dieter Eicke, 820–840. Zurich: Kindler Verlag, 1977.

Kirsch, James. "Reflections at Age Eighty-Four." In *A Modern Jew in Search of a Soul*, edited by Spiegelmann and Jacobson, 147–155. Phoenix: Falcon Press, 1986.

Klemperer, Klemens von. *Germany's New Conservatism: Its History and Dilemma in the Twentieth Century.* Princeton: Princeton University Press, 1968.

Kölbing, M.H. "Wie Karl Gustav Jung Basler Professor Würde." *Basler Nachrichten Sonntagsblatt* (September 26, 1954): 4–8.

Kramer, Fritz W. "Die Aktualitaet der Exotischer. Der Fall der 'Kulturmorphologie' von Frobenius und Jensen." In *Die Restauration der Götter: Antike Relgion und Neo-Paganismus,Fritz W. Kramer*, edited by Richard Faber und Renate Schlesier, 258–270. Wurzburg: Konigshausen und Neumann, 1986.

Kroth, Jerry. *Omens and Oracles.* New York: Praeger, 1992.

Kuspit, Donald. "A Mighty Metaphor: The Analogy of Archeology and Psychoanalysis." In *Sigmund Freud and Art*, edited by Lynn Gamwell and Richard Ellis, 133–151. Albany and London: State University of New York and the Freud Museum of London, 1989.

Labor, Adam. *Hitler's Secret Bankers: The Myth of Swiss Neutrality During the Holocaust.* Seacacus, NJ: Birch Lane Press, 1997.

Lammers, Ann, and Adrian Cunningham, ed. *The Jung-White Letters.* London and New York: Routledge [Philemon Series], 2007.

Langer, Walter C. *The Mind of Adolf Hitler: the Secret Wartime Report.* New York: New American Library, 1972.

Laqueur, Walter. *Young Germany: A History of the German Youth Movement.* New Brunswick: Transaction, 1984.

Large, David Clay. *Where Ghosts Walked: Munich's Road to the Third Reich.* New York: Norton, 1997.

Lears, T.J. Jackson. *No Place of Grace: Antimodernism and the Transformation of American Culture, 1880–1920.* Chicago: University of Chicago Press, 1994.

Leitz, Christian. *Sympathy for the Devil: Neutral Europe and Nazi Germany in World War II.* New York: New York University Press, 2001.

Lenoir, Timothy. *The Strategy of Life.* Chicago: University of Chicago Press, 1982.

Leo Frobenius zum 60th Geburtstag. Leipzig: Köhler und Amelang, 1933.

Lèon Poliakov. *The Aryan Myth.* New York: Barnes and Noble, 1996.

Lexikon der Konservatismus. Graz-Stuttgart: Stocker, 1996.

Lexikon zur Parteigeschichte: Vol. 2. Berlin; Pahl-Pugenstein, 1984.

Lieberman, E. James. *Acts of Will.* New York: Free Press, 1985.

Liebeschütz, Hans. "Das Judentum im Geschichtsbild Jacob Burckhardts." *Yearbook of the Leo Baeck Institute*, 1959.

Ljunggren, Magnus. *The Russian Mephisto: A Study of the Life and Work of Emilii Medtner.* Stockholm: Acta Universitatis Stockholmiensis, Stockholm Studies in Russian Literature, 1994.

Lockot, Regine. *Erinnern und Durcharbeiten Zur Geschichte der Psychoanalyse und Psychotherapie im Nationalsozialismus.* Frankfurt am Main: Fischer Taschenbuch Verlag, 1985.

Lohman, Hans Martin, ed. *Psychoanalyse und Nationalsozialismus.* Frankfurt am Main: Fischer Verlag, 1984.

Lougee, Robert. *Paul De Lagarde 1827–1891: A Study in Radical Conservatism in Germany.* Cambridge: Harvard University Press, 1962.

———. "German Romanticism and Political Thought." In *The European Past: Vol. 2.* New York: Macmillan, 1964.

Luchsinger, Katrin. "Ein unveroffentlichtes Bild von C.G. Jung: 'Landschaft mit Nebelmeer.'" *Analytische Psychologie* 39 (1987): 298–302.

Maeder, Alphons. *Ferdinand Hodler: Eine Skizze.* Zurich: Rascher, 1916.

Maidenbaum, Aryeh. "Lingering Shadows: A Personal Perspective." In *Lingering Shadows: Jungians, Freudians, and Anti-Semitism,* edited by Aryeh Maidenbaum and Stephen Martin, 297–300. Boston: Shambhala, 1991.

Makela, Maria. *The Munich Secession: Art and Artists in Turn-of-the-Century Munich.* Princeton: Princeton University Press, 1990.

Marchand, Suzanne. *Down from Olympus: Archaeology and Philhellenism in Germany, 1750–1970.* Princeton: Princeton University Press, 1996.

———. *German Orientalism in the Age of Empire: Religion, Race and Scholarship.* New York: Cambridge University Press, 2009.

Marcuse, Herbert. *Eros and Civilization.* New York: Vintage, 1962.

Mattioli, Aram. "Jacob Burckhardts Antisemitismus. Eine Neuinterpretation aus mentalitätsgeschichtlicher Sicht." *Schweizerische Zeitschrift für Geschichte* 49 (1999): 496–529.

McDermott, Dana Sue. "Creativity in the Theatre: Robert Edmond Jones and C.G. Jung." *Theatre Journal* (May 1984): 213–230.

McDougall, William. *The Group Mind.* New York: Putnam's, 1920.

———. *Is America Safe for Democracy?* New York: Scribner's, 1921.

———. *Outline of Abnormal Psychology.* New York: Scribner's, 1926.

———. *Modern Materialism and Emergent Evolution.* London: Metheun, 1934.

McGuire, William, ed. *The Freud/Jung Letters: The Correspondence between Sigmund Freud and C. G. Jung.* Princeton: Princeton University Press, 1974.

———. *Bollingen: An Adventure in Collecting the Past.* Princeton: Princeton University Press, 1982.

———. "Firm Affinities." *Journal of Analytical Psychology* 40:3 (1995): 301–326.

Means, Paul. *Things that are Caesar's.* New York: Round Table Press, 1935.

Mellon, Paul. *Reflections in a Silver Spoon: A Memoir.* New York: William Morrow, 1992.

Mertz-Rychner, Claudia, ed. *Carl J. Burckhardt-Max Rychner Briefe 1926–1965.* Frankfurt am Main: Fischer Verlag, 1970.

Metcalf, Philip. *1933.* Sag Harbor, NY: Permanent Press, 1988.

Metzner, Ralph. *The Well of Remembrance.* Boston: Shambhala, 1994.

Mitchell, Mike. *Vivo: The Life of Gustav Meyrink.* Sawtry, U.K.: Dedalus, 2008.

Mohler, Armin. *Die Konservative Revolution in Deutschland.* Darmstadt: Wissenschaftliche Buchgesellschaft, 1994.

Mosse, George. *The Origins of German Ideology.* New York: Grosset and Dunlop, 1964.

———. *Toward the Final Solution.* New York: Howard Fertig, 1978.

———. *Germans and Jews.* Detroit: Wayne State University Press, 1987.

Müller, Guido. *Deutsch-Franzosische Gesellschaftsbeziehungen nach dem Ersten Weltkrieg.* Aachen: Habilitationsschrift, 1997.

———. *Europäische Gesellschaftsbeziehungen nach dem Ersten Weltkrieg.* München: R. Oldenbourg Verlag, 2005.

Muller, Jerry. *The Other God that Failed: Hans Freyer and the Deradicalization of German Conservatism.* Princeton: Princeton University Press, 1987.

Müller-Hill, Benno. *Murderous Science.* Oxford: Oxford University Press, 1988.

Nagy, Marilyn. *Philosophical Issues in the Psychology of C.G. Jung.* Albany: State University of New York Press, 1991.

Neumann, Franz. *Behemoth.* New York: Harper and Row, 1966.

Nietzsche, Friedrich. *The Portable Nietzsche.* New York: Viking Press, 1970.

Noll, Richard. *The Jung Cult.* Princeton: Princeton University Press, 1994.

———. *The Aryan Christ.* New York: Random House, 1997.

Oeri, Albert. "Some Youthful Memories." In *C.G. Jung Speaking,* edited by William McGuire and R.F.C. Hull, 3–10. Princeton: Princeton University Press, 1977.

Ostow, Mortimer. "Letter to the Editor." *International Review of Psychoanalysis* 4 (1977): 377.

Otto, Rudolph. *The Idea of the Holy,* John W. Harvey trans. Oxford: Oxford University Press, 1970.

Persico, Joseph. *Piercing the Reich.* New York: Viking, 1979.

Philipson, Morris. *Outline of a Jungian Aesthetics.* Evanston: Northwestern University Press, 1963.

Poewe, Karla. *New Religions and the Nazis.* New York: Routledge, 2006.

Pogson, Beryl. *Maurice Nicoll: A Portrait.* New York: Fourth Way Books, 1987.

Portmann, Adolf. "Jung's Biology Professor: Some Recollections." *Spring 1976,* 148–154.

Primal Vision, Selected Writings of Gottfried Benn. London: Marion Boyars, 1976.

Pringle, Heather. *The Master Plan: Hitler's Scholars and the Holocaust.* New York: Hyperion, 2006.

Prochnik, George. *Putnam Camp.* New York: Other Press, 2006.

Proctor, Robert. *Racial Hygiene: Medicine under the Nazis.* Cambridge: Harvard University Press, 1988.

———. *The Nazi War on Cancer.* Princeton: Princeton University Press, 1999.

Progoff, Ira. *Jung's Psychology and Its Social Meaning.* New York: Grove Press, 1953.

Rauschning, Hermann. *Hitler Speaks.* London: Thornton, Butterworth, 1940.

Redlich, Fritz. *Hitler, Diagnosis of a Destructive Prophet.* New York: Oxford University Press, 1999.

Reid, Jane Cabot, ed. *Jung, My Mother, and I: The Analytic Diaries of Catherine Rush Cabot.* Zurich: Daimon Verlag, 2001.

Rice, James. "Russian Stereotypes in the Freud-Jung Correspondence." *Slavic Review* 41 (Spring 1982): 19–34.

Ringer, Fritz. *The Decline of the German Mandarins.* Hanover, NH : University Press of New England, 1990.

Roazen, Paul. *Freud and His Followers.* New York: Knopf, 1976.

Roback, A.A. *Psychodrama: A Mental Outlook and Analysis.* Cambridge, MA: Sci-Art Publishers, 1942.

Robinson, Forrest. *Love's Story Told: A Life of Henry Murray.* Cambridge: Harvard University Press, 1992.

Rohan, Karl Anton. *Die Geistige Problem Europa von Heute.* Vienna: Verlag der Wila, 1922.
Rohrbach, H. *Kleine Einführung in die Charakterkunde.* Leipzig: B.G. Tuebner Verlag, 1934.
Rosenbaum, Ron. *Explaining Hitler.* New York: Random House, 1998.
Rosenblum, Robert, Maryanne Stevens, and Ann Dumas. *1900: Art at the Crossroads.* New York: Solomon Guggenheim Museum, 2000.
Rosenzweig, Saul. *The Historic Expedition to America (1909): Freud, Jung, and Hall the Kingmaker.* St. Louis: Rana House, 1994.
Ryce-Menuhin, Joel. "The Symbolists in Art and Literature: their Relationship to Jung's Analytical Psychology." *Harvest* 41:1 (1995): 54–62.
Santayana, George. *Egotism in German Philosophy.* New York: Scribner's, n.d.
Sauerbruch, Ferdinand. *A Surgeon's Life.* London: Andre Deutsch, 1953.
Saunders, Frances Stonor. *The Cultural Cold War.* New York: The New Press, 1999.
Schäfer, Renate. "Geschichte des Wortes 'zersetzen.'" *Zeitschrift für Deutsche Wortforschung* 18 (1962): 40–80.
Scheijen, Sjeng, ed. *Working for Diaghilev.* The Netherlands: Groninger Museum, 2004.
Scheler, Max. *Man's Place in Nature.* New York: Noonday Press, 1971.
Schleich, Carl Ludwig. *Those Were Good Days!.* London: George Allen and Unwin, 1935.
Schmitz, Oscar A.H. *Dämon Welt.* Munich: Georg Müller, 1926.
———. *Ergo Sum.* Munich: Georg Müller, 1927.
———. *Bürgerliche Boheme.* Bonn: Weidle Verlag, 1998.
Schoenl, William. *C.G. Jung: His Friendship with Mary Mellon and J.B. Priestley.* Wilamette, IL: Chiron, 1998.
Scholem, Gershom, and Theodor W. Adorno, eds. *The Correspondence of Walter Benjamin.* Chicago: University of Chicago Press, 1994.
Schom, Alan. *Survey of Nazi and Pro-Nazi Groups in Switzerland 1930–1945.* Los Angeles: The Simon Wiesenthal Center, 1998.
Schwabe, Klaus. "Zur politischen Haltung der Deutschen Professoren im Ersten Weltkreig." *Historische Zeitschrift,* 193 (1961): 601–634.
Scott, William B., and Peter M. Rutkoff. *New York Modern.* Baltimore: Johns Hopkins University Press, 1999.
Selesnick, Alexander. *History of Psychiatry.* New York: Harper and Row, 1964.
Serrano, Miguel. *C.G. Jung and Hermann Hesse: A Record of Two Friendships.* New York: Schocken Books, 1966.
———. *Adolf Hitler, el Ultimo Avatara.* Bogota: Editorial Solar, 2000.
Shaeffer, Louis. *O'Neil: Son and Artist.* Boston: Little, Brown, 1973.
Shamdasani, Sonu. "Introduction." In *The Psychology of Kundalini Yoga,* edited by Sonu Shamdasani, xvii–xlvi. Princeton: Princeton University Press, 1996.
———. *Cult Fictions.* London, New York: Routledge, 1998.
———. *Jung and the Making of Modern Psychology.* Cambridge: Cambridge University Press, 2003.
———. *Jung Stripped Bare By His Biographers, Even.* London: Karnac Books, 2005.
Sherry, Jay. "Jung, the Jews, and Hitler." *Spring 1986* 163–175.
———. "The Case of Jung's Alleged Anti-Semitism." In *Lingering Shadows,* edited by Aryeh Maidenbaum and Stephen Martin, 117–132. Boston: Shambhala, 1991.

———. "Jung, Anti-Semitism, and the Weimar Years (1918–1933)." In *Jung and the Shadow of Anti-Semitism*, edited by Aryeh Maidenbaum, 21–38. Berwick, ME: Nicolas-Hays, 2002.

———. "A Pictorial Guide to the Red Book." In *ARAS Connections: Image and Archetype*. 2010, Issue 1. <info@aras.org>

Shirer, William. *Rise and Fall of the Third Reich*. New York: Simon and Schuster, 1960.

Simpson, Christopher. *Splendid Blond Beast*. Monroe, ME: Common Courage Press, 1995.

Skinner, Cornelia Otis. *Madame Sarah*. Boston: Houghton Mifflin, 1967.

Slossen, Edwin. *Major Prophets of Today*. Boston: Little, Brown, 1914.

Sluga, Hans. *Heidegger's Crisis: Philosophy and Politics in Nazi Germany*. Cambridge: Harvard University Press, 1993.

Smith, Woodruff. *Politics and the Sciences of Culture in Germany 1840–1920*. Oxford: Oxford University Press, 1991.

Solovieva, Olga. "'Bizarre Epik des Augenblicks': Gottfried Benn's 'Answer to the Literary Emigrants' in the Context of His Early Prose." *German Studies Review* (Volume 33, No. 1, 2010): 119–140.

Stackelberg, Roderick. *Idealism Debased: from Völkisch Ideology to National Socialism*. Kent, OH: Kent State University Press, 1981.

Stanislawski, Michael. *Zionism and the Fin de Siècle, Cosmopolitanism and Nationalism from Nordau to Jabotinsky*. Berkeley: University of California Press, 2001.

Stark, Gary. *Entrepreneurs of Ideology: Neoconservative Publishers in Germany, 1890–1933*. Chapel Hill: The University of North Carolina Press, 1981.

Steding, Christoph. *Das Reich und die Krankheit der Europäischen Kultur*. Hamburg: Hanseatische Verlag, 1938.

Stein Ugelvik Larsen, Bernt Hagtvet, and Jan Peter Myklebust, ed. *Who Were the Fascists: Social Roots of European Fascism*. Bergen, Oslo, Tromsph [o/]: Universitets Forlaget, 1980.

Steiner, Gustav. "Gustav Steiner, 'Erinnerungen an Carl Gustav Jung aus der Studentenzeit' in the Basler Stadtbuch Gustav Erinnerungen an Carl Gustav Jung aus der Studentenzeit." In *Basler Stadtbuch*, 117–163. Basel: Verlag Helbing & Lichtenhahn, 1965.

Stern, Fritz. *The Politics of Cultural Despair*. Berkeley: University of California Press, 1974.

Sternhell, Zeev. *Neither Right nor Left: Fascist Ideology in France*. Princeton: Princeton University Press, 1986.

———. *The Birth of Fascist Ideology*. Princeton: Princeton University Press, 1994.

Struve, Walter. *Elites Against Democracy*. Princeton: Princeton University Press, 1973.

Sufranski, Rüdiger. *Martin Heidegger, Between Good and Evil*. Cambridge: Harvard University Press, 1998.

Surette, Leon. *The Birth of Modernism: Ezra Pound, T. S. Eliot, W. B. Yeats, and the Occult*. Montreal and Kingston: McGill-Queen's University Press, 1993.

Switzerland, National Socialism, and the Second World War. Swiss Federal Historical Commission, Zurich: Pendo Verlag, 2002.

Taylor, Eugene. *Shadow Culture*. Washington, D.C.: Counterpoint, 1999.

Thomas Mann Diaries 1918–39. London: Andre Deutsch, 1983.

Toland, John. *Adolf Hitler*. New York: Doubleday, 1976.
Transzendenz als Erfarhung, Beitrag und Widerhall: Festschrift zum 70. Geburtstag von Graf Dürckheim. Weilheim/OBB: Otto Wilhelm Barth-Verlag, 1966.
Treasures from Basel. Maastricht, The Netherlands: The European Fine Art Foundation, 1995.
Treitel, Corinna. *A Science for the Soul, Occultism and the Genesis of the German Modern*. Baltimore: The Johns Hopkins University Press, 2004.
Valery, Paul. *CW 10: History and Politics [Bollingen Series XLV]*. New York: Pantheon, 1962.
Von der Tann, Matthias, and Arvid Erlenmeyer, ed. *C. G. Jung und der Nationalsozialismus*. Berlin: Deustschen Gesellschaft für Analytische Psychologie, 1993.
Von Hassell, Agostino, and Sigrid MacRae. *Alliance of Enemies*. New York: Thomas Dunne Books, 2006.
Waite, Robert. *The Psychopathic God: Adolf Hitler*. New York: Basic Books, 1977.
Walser, Hans. "An Early Psychoanalytic Tragedy—J.J. Honegger and the Beginnings of Training Analysis." *Spring 1974* 243–255.
Warlick, M.E. *Max Ernst and Alchemy*. Austin: University of Texas Press, 2001.
Webb, James. *The Occult Establishment*. Glasgow: Richard Drew Publishing, 1981.
Wehr, Gerhard. *An Illustrated Biography of C.G. Jung*. Boston and Shaftesbury: Shambhala, 1989.
Weinrich, Max. *Hitler's Professors*. New York: Yiddish Scientific Institute, 1991.
Weiss, Peg. *Kandinsky in Munich: The Formative Jugendstil Years*. Princeton: Princeton University Press, 1979.
———. *Kandinsky in Munich: 1896–1914*. New York: The Solomon Guggenheim Museum, 1982.
Weiss, Sheila. *Race Hygiene and National Efficiency: The Eugenics of Wilhelm Schallmayer*. Berkekley: University of California Press, 1987.
Werner, Meike. "Provincial Modernism: Jena as a Publishing Program." *Germanic Review* 76:4 (Fall 2001): 319–334.
West, Shearer. *Fin de Siècle: Art and Society in an Age of Uncertainty*. Woodstock: The Overlook Press, 1994.
Whitford, Frank. *Gustav Klimt*. New York: Crescent Books, 1994.
Williamson, George S. *The Longing for Myth in Germany*. Chicago: University of Chicago Press, 2004.
Wipperman, Wolfgang, and Michael Burleigh. *The Racial State: Germany 1933–1945*. New York: Cambridge University Press, 1991.
Wittels, Fritz. *Sigmund Freud: His Personality, His Teaching, and His Friendship*. New York: Dodd, Mead, & Co., 1924.
Ziolkowski, Theodore. *The View from the Tower*. Princeton: Princeton University Press, 1998.
Zofingia Report 1821–1902. Basel: Buchdruckerei Kreis, 1902.
Zöllner, Johann. *Transcendental Physics*. Montana: Kessinger, 1881 (reprint, n.d.).

Index

CPSIA information can be obtained at www.ICGtesting.com
230444LV00002B/1/P

9 780230 102965